D1345998

Ancient Battles
for
Wargamers

Charles Grant

Ancient Battles for Wargamers

Model and Allied Publications
Argus Books Limited
14 St James Road, Watford
Hertfordshire, England

First published 1977

ISBN 0 85242 553 8

Printed in Great Britain
by W & J Mackay Limited, Chatham

To Nell
In consideration of her efficient operation of the commissariat during many a wargame.

CONTENTS

LIST OF COLOUR PLATES

FOREWORD

THIS IS A BOOK for wargamers, primarily those operating in the ancient period, but also, it is hoped, for the more general reader who may find some interest in the colourful history of the very distant past. Basically, the volume consists of descriptions of four of the most significant battles of those far-off days, to which are joined detailed accounts of wargames which attempted—with what success the reader may judge—to simulate the course of those antique conflicts. To these has been added a considerable amount of material on certain of the more notable troops who have some claim to fame in this most fascinating period, and of whom a knowledge may assist in the understanding of the military art of the time.

For all this, it is a pleasure for me to acknowledge my debt to Alec Gee, Editor of *Military Modelling* and *Battle*, in whose magazines the original material of this book first appeared. My gratitude to him is very great, as indeed it is to his Assistant Editor, Mrs Pat McCarthy. I owe much, also, to those members of the wargaming group whose names figure in the battle reports, and without whose enthusiasm and participation the wargames described could not have taken place. We—in ancient 'Dubris'—although all members of the Society of Ancients, form no recognised club, have neither officers nor constitution, but are simply a number of friends whose similarity of interests has drawn them together over several years, this, I believe, being the best manner in which to form any sort of wargaming association. We have had many splendid wargames and I look forward to more in the future. My debt to them is considerable.

Charles Grant
December 1976

INTRODUCTION

IT HAS FOR SOME time been a fairly staunchly held belief on my part
that wargaming is an end product of a basic interest in the history of
warfare. One does not ordinarily come across a wargamer who one
day out of the blue decided that he wished to wargame; but rather it
will be found that a newcomer to the hobby has already a leaning, no
matter how slight, towards some aspect of the military history of one
period or another, with particular reference to how the contending
forces manoeuvred, what weapons they used, the manner in which
the various arms operated in conjunction, and how the generals
deployed their troops and ordered their battles. It is really but a short
step from this to the thought that one could have done better in the
place of, say, Hannibal, Gustavus, Adolphus or Montgomery, or at
least the belief that one would have gone about things in a different
way. This is the crucial point, then, when the 'historian' becomes
also the wargamer.

It is my contention that the wargamers should make every effort to
achieve historical accuracy in his wargames. While it is perfectly true
that 'playability'—an awkward word—must be considered in any
approach to wargaming, adherence to any such principle should in
no way inhibit a desire for realism and accuracy. In this connection I
can only base my judgement on my own not inconsiderable experience and at the same time crave forgiveness for any suggestion of
immodesty. The simple fact is that I have participated in, umpired, or
merely been an observer at hundreds of wargames and in no case
have I seen one which was improved by a distortion or abandonment
of rules which truly reflected military practice—irrespective of the
historical period involved. Moreover I have never witnessed a game
governed by accurate and realistic rules which was not enjoyed and
which was not productive of excitement and interest. This is not to
say, of course, that there is not a point of balance, where a limit must
be drawn to the minutiae of rulemaking, and here I feel that complexity of wargame rules is not necessarily a synonym for realism. I

recall, with mixed feelings, taking over from a player in a modern game who had to go elsewhere for an appointment or something. The rules used were well known at the time but shall now be nameless. During the couple of hours or so in which I was involved, we—my opponent and I—completed just two moves. The fact that two rule books or codes of cause and effect were used was a contributory factor—once certain facts had been established from Book One, Book Two had to be consulted for further information and directions. It was, all in all, a harrowing experience I should not care to repeat.

The purpose of this preamble is really to indicate my belief that realism is by no means incompatible with enjoyment and that complexity of rules is not essential to that quality. Fortunately, too, rules are not immutable in the wargame and one of its most significant features is that each battle will almost certainly produce some military situation which has not previously been encountered and which necessitates some degree of interpretation before it can be assessed, so that the game can proceed smoothly and without snags of any kind. What must be, I feel, the ultimate test of wargame rules is how they show up when being used for the reconstruction or re-enactment of an actual historical battle, and how, when such an operation has been undertaken, it can be carried through—irrespective of its outcome or whether the result reflects that of its historical prototype—without any gross infraction of military probability. It is the main purpose of this book to illustrate just how such reconstructions are both informative and highly enjoyable, not to say exciting.

It has indeed been written that such 'action replays' in imitation of historical engagements are liable to failure for the reason that, armed with hindsight, players will not make the mistakes made by the actual historical generals and that consequently these refights will not resemble in any way that which they purport to reproduce. This is not so, and one way of overcoming such a difficulty—if indeed it is accepted as such—is to ensure that the players taking part remain in ignorance of the battle they are reconstructing and are thus in the same position—*vis-à-vis* reconnaissance or intelligence information—as their forerunners. This, it has to be said, is not always possible, especially with players who have studied their period and can therefore be expected at least occasionally to identify the armies involved if not the actual conflict. This, however, is not an insuperable obstacle if the players start the battle at a point where they are committed to the objectives that were in the mind of their actual predecessors. This is not really a tremendously difficult thing to achieve and needs but a working knowledge of the course of the battle for it to be put into operation. Obviously, should the situation be set up at too early a stage the engagement may take quite a different course from that which it in fact did and the object of the exercise is nullified. By this it is not intended that the players should slavishly follow the tactics and manoeuvres of history but rather that they should operate within an acceptable and logical tactical framework, having the same objectives in mind.

Here it has to be said that, if practice does not necessarily make perfect, some experience in these matters does help in choosing the situation in the historical battle where it appears most conducive to re-enactment. Happily, over the years I have been fortunate in being able to refight—with eager and highly competent assistance—a wide variety of such wargames, covering a diversity of periods and ranging from the Battle of Kadesh in 1288 BC (which I take the liberty of presenting to the reader in this work) to such a vastly different affair as the Battle of Gettysburg in 1865, by way of the English Civil War, the Seven Years War and the Napoleonic War, and I have to say that none of these has proved more rewarding than those set in what we call the 'ancient period'. There is tremendous attraction in examining and analysing, through the medium of a wargame, the actions and decisions of long dead generals and I cannot conceive of any better way of doing this than by the restaging of the great battles wherein the contemporary military art was demonstrated in all its strengths and weaknesses. I think it is true to say that the amount of identification by players with their real life counterparts is far more apparent in these historical refights than in any other sort of wargame—by no means a bad thing. It is perfectly correct that in competition games with two players ranged against each other there is the undeniable excitement of the rivalry of the two contestants intent on defeating each other, but in the multi-player battles of which I speak there seems to be a very real wish on the part of the participants to emulate or outdo the historical character whose identity has been temporarily assumed. We—and I speak of the players who have participated in the battles I take the opportunity to describe in these pages—reach a far higher peak of excitement and tension than in any other contest and many is the mature and experienced wargamer I have seen so affected. I recall in one such game—a simulation of a certain engagement in Caesar's conquest of Gaul—one section of the battle (it was fought in separate actions, something which is often desirable) involved a couple of Roman legions holding an isolated fort which was being assailed by vastly superior forces of Gallic tribesmen. The Romans knew that they had to hold out for some time but they did not know if, or when, reinforcements would reach them. The Roman 'general' whose name appears more than once in this volume, was in something of a 'state', as wave after wave of Gauls broke against his ramparts and his own ranks became thinner and thinner. His eyes glittered, and perspiration streamed in substantial rivulets down his expressive countenance. I'm pretty sure it is a rare thing to see an ordinarily mature and rational chartered surveyor in such a febrile condition. It was for him an exhausting but enjoyable experience, particularly as his Romans were just able to hold out until the arrival of reinforcements under Caesar himself—just as happened historically, in fact. Not that such behaviour is uncommon—I have seen it evinced by such different characters as a grammar school boy and the captain of a multi-thousand ton cross-Channel vessel.

Having thus, I hope, made the point that these reconstitutions are productive of great fun it remains only for me to make a few suggestions as to how they are organised (the reader will find further hints to that end in each individual battle narrative but these are the basics). I stress that they are but suggestions, for a great deal will depend upon the type of wargaming organisation involved and the willingness of the members to participate. The actual mechanics of such operations are not too complicated and, once a decision has been reached on the battle to be fought, it is easy to get going. Obviously one cannot do much with a small table, so the bigger the better. The greater the playing area the more miniature figures can be deployed, and naturally the larger the forces being moved about the greater will be the feeling of being concerned in a major battle. We are now getting to the nub of the business, presupposing that all the necessary study and research have been carried out and that the requisite data as to terrain, armies engaged, and tactical background are at the finger tips of the organisers. With this the case, as the first step one simply sets out the battlefield with on it the essential topographical features. The ancient period quite often gives us an advantage in this connection, many of the battles being fought over relatively flat countryside, with a few obstacles to negotiate or built up areas, villages, farms and so on to impede movement, such as one would encounter, for instance, in wargames set in the Seven Years War or the American Civil War, to name only two. We do have the occasional hill or river, but it was rare for an ancient set-piece battle to involve the significant use of something like buildings in a true tactical sense, as for instance La Haye Sainte or Hougomont at Waterloo. Indeed for many ancient armies it was a virtual necessity to seek flat ground and we have at least one example of ground being smoothed, by King Darius at Gaugamela to facilitate the movement of his chariots.

So, when the terrain has been organised, making sure that the tactical features—if any—of the original battlefield are reproduced in such a way as to give them the same influence as their historical prototypes, we proceed. In passing, although the terrain can be absolutely flat—as at Carrhae in 53 BC—if a rather more complex affair presents more of an attraction, then what about the defeat of the Roman legions in the Teutoburger Wald in AD 9? This would be a fascinating operation for any wargamer. In any event, we go on to set out the contending armies. Let us suppose—and indeed this does happen in many ancient battles—that the two armies are present in their entirety. Here may I be allowed to sound a note of warning. In these circumstances I eschew any system of proportional representation, that is, giving a fixed ratio of actual troops to a single wargame figure, such as one figure equals twenty men, thirty men or whatever, and then scaling down units proportionately. I do not favour this, and indeed I have not found it works too well. It is far more efficacious to study the historical battle order of the armies, note the frontages covered by the different troop types, and then set down one's wargame figures to cover the proportional frontages. A simple example—if your historical battle shows a Roman line with the central two-thirds occupied by legions, then the corresponding two-thirds on the wargame table are filled by your legionary figures, organised in as many units as are thought best to relate to the 'real thing'. In this, though, it is better, if a choice exists, to reduce the unit members rather than strive too rigidly to reproduce the real ones. Experience indicates that this tends to produce a 'bitty' sort of effect. Again I have no option but to say that the above is the system we have found to be preferable and the remarks are intended to be suggestions and are far from being dogmatic.

In any case, with the armies in position we can make a start. It is important, for it ties in with the setting down of the troops, to devote a few words to the question of timing, that is to say, the choice of the moment, when, with the armies set out at preliminary stage of the action, we can abandon historical restraints and allow the wargame generals to take over. This is quite often relatively easy, when two armies are drawn up formally and, at a given moment, advance against each other, but there are times when some cogitation is necessary to determine the critical point. When Plataea (479 BC) was fought, no little attention had to be paid to ensuring the moment when, with the Greek army falling back, the situation was such as logically to invite a Persian attack. The Battle of the Sambre (57 BC) was another even more difficult; with Caesar's legions scattered about foraging, a point had to be chosen so that the assaulting Gauls, bursting from the forest, had a chance to catch the legions before they could reorganise. In the present volume we shall see that the Battle of Kadesh was the only one to pose such a problem, and with the reader's indulgence I hope to demonstrate how it was resolved.

A final point concerns wargame rules. It is a fact that no given set of these, however good, will cover every possible wargame contingency, and from time to time adaptions—which may indeed become part of the rules permanently—have to be made. Indeed, one excellent feature of the contests described in this book is the opportunity they provided for testing the rules in the most rigorous conditions, and several major changes have resulted from anomalies which have appeared in the course of one or other of these wargames. Nothing is perfect and if improvements can be made then the operations I describe were not in vain. In a word, though, this is subsidiary to the main reason for these reconstitutions, which was in effect to provide the maximum of enjoyment for the participants and certainly those detailed did this and a great deal more.

I THE BATTLE OF KADESH 1288 BC

Hittites v Egyptians

FROM THE EARLIEST TIMES nations in the 'Fertile Crescent' of the Middle East waged primitive war with one another, but in none of these cultures was the military art studied to any great extent, and it is not until the establishment of Egypt as a substantial power that we have the appearance of anything which might be described as a regular or standing army. Egypt has always been something of an international cockpit—much as the Low Countries in Europe many years later—and in the millennia before Christ its people had to struggle almost endlessly against aggressive and covetous neighbours. Their fortunes varied from century to century, sometimes being at a complete nadir, as for instance during the occupation of the Hyksos, which lasted from about 1800 BC to 1580 BC, when they were driven out. This lengthy occupation was not entirely without benefit for the Egyptians, who learned much from their overlords, by them being introduced to the powerful composite bow and the chariot, later to be such a feature of the Pharaonic order of battle. It was under these kings— the Pharaohs—that Egyptian frontiers were consolidated and even extended, Egyptian troops penetrating through Syria as far as the Euphrates. Thus, by 1447 BC, the Kingdom of Egypt—under Thutmose III, who died in that year and who was one of the great warrior Pharaohs—was possibly at its zenith.

It was in their northward thrust through Syria that the Egyptians clashed with a people who were to be their deadly enemies, the Hittites, whose lands lay around the south-eastern portion of what is now Asiatic Turkey. (*Fig. 1.1* gives the positions of the two territories involved.) These Hittites, a vigorous people, carried out continual attacks against their southern neighbours, with some success, but in 1292 BC a new and ambitious Pharaoh ascended the throne of Egypt. This was the famous Ramses II, who took it upon himself to restore the splendour of his kingdom and to assert the power of his arms,

1

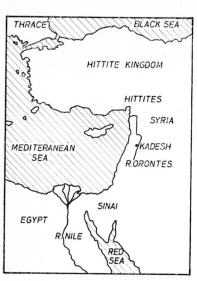

Fig. 1.1   Egypt, Syria and the Hittites   Fig. 1.2   Situation before the battle

particularly *vis-à-vis* the old enemy, the Hittites. One of his prede-cessors, Seti I (1318 to 1299 BC) had, indeed, made a bold attempt to destroy the ascendancy of the Hittites, but had been unable to do so and had in fact left the military situation on the northern frontier of Egypt in an exceedingly parlous state, with a considerable portion of the one-time territorial gains in Syria now in the hands of the Hittites. The latter had done a considerable amount of work in consolidating their advantage in the area and in particular had built up Kadesh on the River Orontes into a near-impregnable citadel (we shall be hearing more of this strong place before long). Ramses was doubtless aware of this and he devoted the first years of his reign to building up a powerful army, the strongest so far seen in Egypt, in preparation for a northwards move. As events were to prove, he needed every man, but it is certainly to his credit that he did much to establish a really well organised and disciplined force, both from the native Egyptians and from a variety of mercenaries recruited from the neighbouring regions. It might, indeed, be to our advantage at this point briefly to consider the army Ramses had at his disposal and at the proper time similarly to look at the enemy he proposed to confront.

What should be noted initially is the fact that, while the mind immediately and inevitably associates the armies of ancient Egypt with the chariot, these vehicles were only a part of a highly organised and—within the limits of the technology of the time—a well balanced force. Up to the era with which we are concerned—that of Ramses II —a high degree of organisation had been developed, with the standing army consisting of 3 divisions. To these a 4th was added by Ramses himself, and they were named, fairly dramatically, the divisions of Amun, Re, Sutehk and Ptah, each apparently numbering some 5,000 troops of all arms, and each being commanded on active service by a royal prince. Alas, no details are on record of the proportion of different types of soldier in each division, but each had its chariot component, plus infantry of various sorts (there were no cavalry units, this, it is presumed, because the horse as it was at that time was not powerful enough to carry men in battle, although it does appear that mounted messengers were employed).

The principal weapon of the Egyptian infantry was the bow, this being especially favoured by the Nubian auxiliaries—from the Sudan —and it was also employed by the chariot crews. The latter consisted of two men—a driver and an archer. Javelins were also carried in racks or cases attached to the sides of the chariot. Armour was known, and worn, in the shape of small rectangular plates of bronze sewn on to a tunic, but I do not think there could have been very many men so accoutred. Officers were more likely to be equipped in this fashion or small bodyguard units or the like. Indeed the clothing worn by the Egyptian rank and file was pretty exiguous and would generally be limited to the ubiquitous loincloth in various styles. Other weapons too were pretty varied—the axe and the 'khepesh' being characteristic, together with pole-axes and maces, the last seemingly carried by officers. The shield, as borne by the infantry, was commonly large and rectangular, with a rounded or pointed top.

In addition to the native Egyptians and the Nubians we have already mentioned, considerable numbers of mercenaries were assembled by Ramses for his campaign—Libyans and Sherdan, for instance. The last named are more than a little fascinating, having a most dramatic appearance, with leather body protection edged and studded with bronze, a round shield, horned helmet, and carrying spear or sword. Their origins are debatable, but they may have emanated from present-day Sardinia or neighbouring regions, and they seem to have been part of the 'Peoples of the Sea'. They were recruited in considerable numbers and may have formed a kind of bodyguard for the Pharaoh.

Any consideration of the tactics employed by the army of ancient
Egypt must be largely conjectural, and we must not overlook the fact
that it is unlikely that it was often faced by any sort of regular force
capable of standing up to the advance of the chariots. These, possibly from an initial position on the wings of the main body of
infantry, would sweep forward on an enemy weakened by archery to
such an extent as to break before the actual contact by the chariots.
The chariots may also have been employed in a skirmishing role,
driving rapidly up to the enemy front and showering it with arrows
before wheeling away. If no break took place it can be presumed that
the Egyptian infantry centre would advance and engage the enemy at
close quarters. I cannot help feeling that their successes of the past
had been against fairly indifferent opposition, and the foe they were
about to face—the Hittites—was to be a very different kettle of fish,
militarily speaking.

In any event, at the beginning of the year 1288 BC, the Pharaoh
Ramses led forth his army, some 20,000 strong, his objective being
the destruction of the Hittite forces in Syria. The march he had to
make was a long one, involving almost the entire length of Palestine,
whence he finally debouched with his troops into the valley of the
Orontes, beyond which lay the north Syrian plain and the fortress of
Kadesh, quite certainly the key to Hittite control of the area. And
there, in the neighbourhood of Kadesh, lay the army of the Hittite
monarch, Mutallu (or Muwatallis, it is a matter of choice). Now there
can be little doubt that the Hittite king was a most astute individual
who, as events were to prove, had had considerable military experience, and he had not by any means wasted the previous few years,
having collected from his own territories and those of his neighbours
and allies a powerful army, numbering some 20,000 men, very much
the same strength as the Egyptians. Many kings contributed their
men to this host, those of Arvad, Carchemish and Aleppo among
them.

As did the army of Ramses, the Hittite force included a high
proportion of chariotry, and numbered many different types of
infantry, armed with spear and sword, while archers were also included. Once again, no cavalry appeared in the Hittite order of battle.
One important fact we must notice is that the Hittite chariot is
believed to have been appreciably heavier than its Egyptian counterpart (which was very light indeed). There is doubt as to the number of
men comprising the Hittite chariot crew—some sources indicating
two and others three men—but whichever is the case the weapons
carrried were not dissimilar to those of the Egyptian charioteers,
being bow and javelin. We seem to have more details of Hittite
chariot numbers (the account of the campaign by Professor J. H.

Breasted in Vol. II of the Cambridge *Ancient History* is a first rate one, but I also rely considerably on articles on the Egyptian and Hittite armies in *Slingshot* by Alan Buttery, as well as on his *Armies of Ancient Egypt and Assyria*) and they seem to have amounted to as many as 3,500 vehicles.

One point which must be noted as it is not without importance is a possible difference between the Egyptian and the Hittite weaponry. That of the former was bronze or copper, but the latter must have had a certain advantage in that at least a proportion of their weapons were of iron. The production of this valuable metal had been known in the Hittite confederacy in 1400 BC and the king of the time, one Shuppilu-lumash, must have used sword blades and spear and axe heads of the harder metal. Apart from the occasional gift of an iron blade to favoured friendly kings—a species of diplomatic gesture—iron weapons must have been virtually entirely in the hands of the Hittites. Finally, as we have said, the Hittite army amounted to 20,000 men, but by no means all were to be engaged in the forthcoming battle.

Having thus briefly examined the two armies, we can proceed, and at this point *Fig. 1.2* will show us the positions of the opposing forces with battle fairly imminent. The Egyptians, with the Amun division in the lead commanded in person by Ramses, has forged ahead, but Re and the others have been left considerably behind and the whole army is spread over many miles of ground. In point of fact its fragmentation may have been appreciably greater than is shown on the map. Ramses' reconnaissance must have been virtually non-existent, for at this stage he had absolutely no idea of the position of the Hittite army to the northwest of Kadesh, which itself was no doubt strongly held. Indeed, earlier in the day when Ramses was at Shabtuna, a couple of local inhabitants on being questioned, stated that the Hittite army was many miles to the north, a tale which the Pharaoh swallowed, hook, line and sinker. In fact, these 'inhabitants' had been specifically sent by Mutallu to lull Ramses into a false sense of security. The Hittite monarch was indeed a devious and crafty individual. Thus completely deceived Ramses pushed on with alacrity, as far as is known making no attempt to scout either ahead or to his flanks, and in so doing left the Re division well behind him.

As Ramses advanced towards Kadesh, the Hittites, keeping the city between them and the Egyptians, crossed the Orontes and moved southwards along its eastern bank. Thus, when in the early afternoon, the Pharoah halted Amun to make camp just beyond Kadesh he was unconscious of the proximity of the enemy, the latter being south-east of Kadesh, directly opposite the flank of the second Egyptian division, Re, toiling northwards to catch up with Amun.

It seems almost incredible that such a movement as carried out by the Hittites should have been made without any sign of the manoeuvre being noted by the Egyptians, but such was the case, and while certainly the lack of reconnaissance on the part of the Egyptians was a contributory factor there must have been more than just this. I feel that there must have been some sort of high ground, possibly a narrow ridge stretching south from Kadesh and along the river bank which was one of the reasons why the Hittite change of position was unobserved. If this be true it goes a long way towards explaining what next transpired, for, concealed from the marching men of Re, who were possibly also surrounded by the dust raised by their own progress, 2,500 Hittite chariots crossed the Orontes probably low at this time) and charged home on the right flank of Re. Surprise was total and the division was utterly shattered, its components scattering in all directions, many men fleeing northwards towards Ramses and Amun, now happily making camp north-west of Kadesh.

A discordant note had been introduced there, however, when a couple of Asiatics were captured, and promptly being 'put to the question', at once divulged the exact location of King Mutallu's army. Pretty well simultaneously with this revelation, it appears, the fugitives from Re came rushing into the camp, with the great mass of Hittite chariots thundering in pursuit. Amun was caught up in the rout and willy-nilly went off to the north along which much of Re. Ramses' position was precarious in the extreme, but his bodyguard— possibly the Sherdan mercenaries—rallied round and for the moment he was safe, although the Hittite chariots were deploying with the intention of enveloping the few Egyptians still disposed to fight. Gathering his guards, Ramses' first intention was to cut his way through the enemy towards the two divisions still far away to the south, but his battle eye did not deceive him and, noting that the Hittite chariots were less densely arrayed on their right—that is, near the Orontes—he and his men hurled themselves furiously at this section of the Hittite line. Such was the fury of the onslaught that the Hittites facing him were flung back, many of them indeed into the river. Among those so brusquely treated were Mutallu's own brother, the chief of his bodyguard and no less a personage than the King of Aleppo, who was recovered from the waters in a sorry state, being brought back to consciousness with some difficulty.

In spite of this success, however, Ramses and his men were sorely pressed, but fortunately for them, the Hittites were diverted by the spectacle of the vast quantities of booty lying around, and immediately fell to plundering the Egyptian camp. However, at this point another force put in an appearance, coming directly from the west. This was the Na'arun—the 'recruits'—and whence they came and what they were is impossible to say. They might have come up from some garrison on the coast, but this seems less likely than the other two possibilities, one, that they formed a flanking force previously sent out by Ramses to protect his left, or two, that they were a part of the stricken Re division now rallied from the flight. Whatever they were, they did much to restore the situation in favour of the Egyptians, for they took the Hittites looting the camp by surprise and dispersed them in all directions. In fact, Mutallu had to send in a further wave of 1,000 chariots but he remained himself on the east bank with a powerful body of some 8,000 infantry, who took no part in the fighting. A very confused combat followed with Ramses putting in attack after attack with the Na'arun, his guard, and some stragglers now coming back to rejoin his standard. At last, too, the Ptah division came up, and although certainly very exhausted by its long march, joined Ramses in driving back the Hittites into Kadesh with heavy losses, night falling with the enemy safe in the city and presumably still on the east bank of the Orontes.

Thus ended the battle. Losses were heavy on both sides and although, tactically, the Egyptians may claim a victory, strategically this was not so—despite their later statements—as they retreated south without more ado, leaving the Hittites in possession of Kadesh, the taking of which had been the Pharaoh's objective.

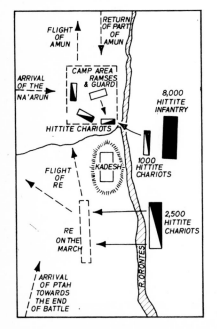

Fig. 1.3 Elements of the battle

The wargame refight

Our reconstitution of the Battle of Kadesh was a fascinating example of the art (or is it craft?) of turning an historical engagement into a wargame, containing, as this battle did, a most satisfying number of unknown factors. A point which cannot be over-emphasised is the importance—nay, the necessity—of retaining the historical elements of the occasion and at the same time—for after all wargamers are human (or most of us are)—of maintaining the all-essential balance which gives each side a fairly reasonable chance of achieving victory within the framework provided by history. This is a feature which, for ordinary purposes, eliminates from possibility of consideration any number of really fascinating battles on the simple grounds that the opposing forces are numerically so disproportionate as to make the result a foregone conclusion within the limits imposed by any acceptable set of wargame rules. After all, enthusiastic as one may be in studying or analysing some particular encounter, it is indeed an

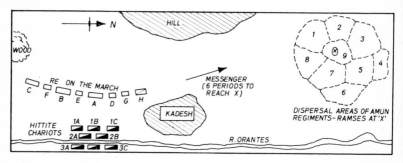

Fig. 1.4   Overall picture of the battlefield—Period One

altruist who will gladly accept the role of a general whose defeat is virtually inevitable.

No problem can be found in the case of Kadesh with any disparity in the numbers of the opposing armies, each side deploying, as we have noted, some 20,000 men, although the proportions of the varying arms in each cannot readily be established. What does merit a little consideration is the determination of the point in the proceedings at which we are to set out our troops ready for the actual wargame to commence. Normally, this would be at a fairly early stage of the action before any significant manoeuvres have begun, but for Kadesh to be fought in the spirit of the event as it occurred, it was thought that a time should be chosen such as would ensure that the players were more or less committed to follow the main stream of historical events. Otherwise, armed with hindsight, they might be able completely to change the course of the action to such an extent as to make the wargame resemble in no way that which they were trying to recreate. Ideally, this could be obviated by making certain that the players had no information as to what was taking place other than that which was available to the contemporary generals. As was pointed out in the Introduction, this is not always possible, especially with wargamers who have read up their period with some diligence and who might well deduce, when presented with a certain situation and terrain, just which battle they were to fight. In the case of the Battle of Kadesh, the number of players involved meant that only a proportion had the necessary knowledge, and even they were not fully cognisant of just how the 'controlling authority'—for want of a better term—had decided that the game would proceed.

Holding firmly then to the precept that the wargame should begin with the players well committed to action which, initially at least, would cause them to follow historical precedent, the battlefield and the action were examined. It seemed hardly practicable to restate the hidden switch of the Hittite army from north-west to south-east of Kadesh, for were the player taking the role of Ramses to observe this—and I cannot conceive of any player's marching blithely on without sending out reconnaissances—he would without question have halted Amun and waited the arrival of Re before progressing further. So, following the course of the battle, prepared to 'freeze' the action at a suitable point, we came to the moment when the Hittite chariots, having crossed the Orontes, are about to fling themselves upon the flank of the marching Re division. This seemed much more promising, and without more ado, the stage was set. Again we have to pass briefly over the ensuing discussion but after some debate it was decided that the battle area would have to contain the terrain where Ramses and Amun were making camp at the time of the attack on Re. This entailed a resetting of the wargames table, ordinarily two 7-ft × 4½-ft sections butted together to provide a 9-ft × 7-ft playing area, but which, for Kadesh, we disposed longitudinally to provide a table 14 ft × 4½ ft. Upon this the requisite terrain features as shown in *Fig. 1.4* were disposed, and in the position shown was Re division on the northwards march.

Now, when one is faced with a terrain, as we are at this stage, there is no need overly to concern oneself with the ratio between wargame figures and actual numerical strengths; one simply fills the area covered by an army—or a division in this case—with wargame figures. As has already been stressed, no information is available on the makeup of the Egyptian divisions, so what was set out was chosen fairly arbitrarily, but bearing in mind what appeared to be reasonable. Thus, in the positions shown in *Fig. 1.4*, Re was organised as shown in table below.

The Egyptian Army

A	Karnak infantry (spear and shield)	40
B	Tharn infantry (spear and shield)	40
C	Carchemish infantry (spear and shield)	40
D	Ka light archer regiment	20
E	Koth light archer regiment	20
F	Kush light archer regiment	20
G	Amarna light javelin regiment	20
H	The Royal Chariot Squadron	8
	Total	208

These regiments—their names are fictional, but in default of historical nomenclature, I make no excuses for my partiality for named units—can be identified on the map by the letters, and a little addition will show that the division of Re numbered 200 figures and 8 chariots. The proportion of spear-armed medium infantry to archers and javelin men seemed to be an appropriate one but again I have to admit that the composition of the division is strictly conjectural. The next stage in the build-up was to locate the first waves of Hittite chariots and to estimate their numbers. To this end 5 chariot units each of 8 vehicles were taken to represent the initial attacking force of 2,500, from which it can be deduced that with 8 vehicles with Re represented 500. That the chariot strength of an Egyptian division was 500 is again a point open to discussion, but for want of evidence we have to accept it. So, 40 chariots made up the original Hittite commitment, the second wave of 1,000 sent in by Mutallu being, in our scale, two more units, but to this was added a third, just for good measure, one might say, so that the follow-up force amounted to 1,500, not 1,000—just a little bit of juggling, of course. In practical wargame terms, a total of 64 chariots were ready to be launched against Re. It must also be admitted, in self-conscious parenthesis, that many of these vehicles had never operated either near the Nile nor the Orontes, and would have been more familiar on the banks of the Thames or the Jhelum, but their function was in no way impaired by their being so far from home.

The next point to be decided was the point reached by the Hittite chariots when they were first observed by the Egyptians. It seems from the account of the battle that they were assailed really before they knew what was happening, but this did not seem proper. It was unlikely that absolutely no warning was received by Re—its

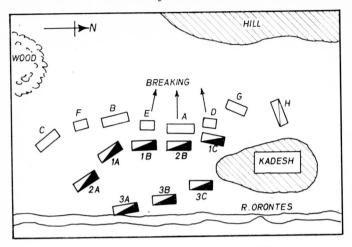

Fig. 1.5   Details of the fighting—Period Six

regiments must at least have the chance to brace themselves a little and for such a mass to approach to such proximity was, I think, a tremendous piece of luck for the Hittites, such as would not ordinarily be repeated. Thus, to give the Egyptians a very slight fighting chance, the Hittites were place as shown in *Fig. 1.4*, ready to charge, but at such a distance that the Egyptians were able to face the attack, but had no time to deploy in any way.

So far, so good. Let us turn our attention to the Amun division, again shown on *Fig. 1.4*, the dotted enclosures being the extent of each unit's dispersal, the figures scattered about in foraging and such like, Pharaoh himself being at Point 'X'. There is no need to give the details of Amun's composition, as much of it played no part in the action. However, it seemed not unreasonable, again with playability in mind, to ensure that Amun would not 'sit out' the battle without doing something, so it was enacted that, on the Hittite charge, a messenger would be despatched from the Royal Squadron of chariots to bring the dire news to Ramses. Travelling at chariot speed, it was calculated that the messenger would arrive in about six periods. Further, if in the messenger's progress towards the monarch, any Egyptian unit was passed, it was deemed that this would alert it to the state of emergency and it would start to rally to its standard, although no move could be undertaken without the Pharaoh's signal. On this point, it must be said here that, wearing the double crown of Egypt for the occasion was Ian Osborn, who, as the latter-day Ramses, had as First Minister Martin Dice. Carrying on with the 'dramatis personae', Mutallu, commanding the second wave of Hittite chariots was Peter Sheppard, the first five units being, for no good reason, under the King of Aleppo, in the person of Andrew Green. After some consideration, the command of Re was divided into four—Alan Angus took the Royal Squadron and the javelins, Rob Waldren's command was units D and A, Ray McGarry had E and B while Gregory Perry had F and C, thus providing three infantry brigades each of an infantry and an archer unit.

And so to battle, with the Egyptians being suddenly surprised by the presence of the Hittite chariots on Period One (see *Fig. 1.4*). They had no chance to shoot at this point, although the Hittites, in their approach, did shoot, causing a few casualties. Reaction tests all along the line, however, showed that the Pharaoh's men were in good heart, and during the charge which took place in the following period, while the archers successfuly evaded the crunch, Karnak infantry actually pushed back the contacting Hittite chariot unit (1B), destroying one chariot in the process. As a result of all this the Hittites had to rally back, giving the Egyptians something of a breathing space, during which the Royal Chariots wheeled to face the enemy and the messenger went bombing off towards Amun to seek assistance and impart tidings of the attack. During the second and third periods, the Kushite archers moved forward to bring the Hittite masses under archery fire and the Carchemish infantry deployed into line, preparatory, it seemed, to an advance against the Hittite left flank. Meantime a charge against the Amarna javelins by Hittite unit 2B had been nimbly evaded, momentarily leaving the latter exposed to a flank attack by the Royal Squadron, but in view of the threatening presence of the now rallied Hittite front line, their commander, doubtless wisely, decided to wait and fight, if not another day, another period.

The next two periods saw the original Hittite attacking squadrons again launched against the Egyptians, with only nominal success however, the Infantry hanging on grimly with the chariots being unable to inflict the necessary casualties to enforcd a 'break' or even a 'push back'. The 40 strong, spear armed infantry units were particularly effective. Finally, Period Six saw the first major Hittite success, Karnak infantry being broken, as well as both Ka and Koth archers. However, simultaneously with this, the Egyptian messenger reached the Pharaoh in the midst of Amun division and at once the signal to his men went forth (*Fig. 1.5* shows details of the actual fighting at this time). King Mutallu is leading his second wave of chariots across the Orontes and a very considerable gap has been punched in Re, much later than Mutallu had hoped, and indeed at the cost of six chariots.

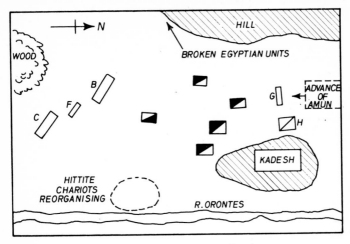

Fig. 1.6 Situation at end of battle

However, with these units in rout, reaction tests were the order of the day and both the Carchemish and the Tharn regiments were unlucky in having a 'halt two periods' enactment. In no little jubilation Hittite unit 1A—heavy chariots—went charging into Tharn but was promptly driven off in rout, much to the discomfiture of the King of Aleppo, whose elite unit it was, no less than three chariots being lost in the debacle. By this time, King Mutallu's additional three units were up, and unit 1A was able to rally in a couple of periods. Carchemish and Kush had taken up a species of defensive position with their backs to the wood. The situation was becoming a trifle fraught for the Egyptians and the Royal Squadron of Chariots, with Amarna javelins in support finally advanced, to be engaged by Hittite unit 3C, pushed back, then broken, with the loss of four chariots (this happening in Periods Eleven and Twelve). Amun was still some distance away (indeed it was only on Period Eight that the Pharaoh's men had got under way), but on the other hand, continuous archery from the Kushites had accounted for several Hittite chariots, and indeed two of their squadrons had been forced back by bad reactions and were reorganising near the Orontes. Joined by the remnants of Tharn infantry, which had repulsed several charges and was consequently much depleted, Carchemish and Kush now posed a substantial threat against any Hittite attempt to make an all-out attack against the northern part of Re, and furthermore, on Period Thirteen, Amun's leading regiment appeared north-west of Kadesh, giving the Hittites occasion to think hard. The result of this cogitation was an immediate regrouping of all available chariots to face the relics of the Royal Squadron and the javelins, close to whose rear were the leading elements of the Amun division, this being the situation at Period Fifteen (see *Fig. 1.6*).

It had been decided that, as the battle began in the afternoon, fifteen periods would suffice to represent fighting time and now night had fallen, so we have to consider the position for the benefit of the participants and to decide on what would be the natural consequences of the fighting as it had progressed. First, due of course to the one period warning given to Re, the initial onslaught was far from being as destructive as was the original, and the Egyptians were able to make a very respectable fight of it—obviously chariots charging home on spear-armed and shielded infantry are not as effective as might have been thought, and of course they are highly vulnerable to archery. When nightfall occurred then, the game situation was that Amun was about to intervene from the north (The Na'arun, of course, did not put in an appearance as they were scheduled to do had the fighting moved in that direction) and Ptah division was approaching from the south. On the other hand, the Hittites had lost 22 chariots, although this left them with 42 out of the original 64 and there was the vast body of infantry—the 8,000 we have spoken of—on the east bank of the Orontes. To be honest, the action of the battle was governed in a more or less off-the-cuff way, but at one time it was debated as to whether this infantry should be brought across the river. The idea was rejected as being likely to alter the battle in a completely unreal way.

In brief, then, our council of war determined that, with Egyptians on both sides, the Hittite chariots would either move into Kadesh, itself strongly held, or join the infantry beyond the river, while the Egyptians would follow history and retreat to the south. This they did. In retrospect, it was a highly enjoyable wargame, giving as much delight in the historical build up as in the actual fighting—a most satisfying operation all round.

A Hittite infantryman with typical arms and equipment

II THE HOPLITE

It may well be that I am sticking my neck out more than a little in saying that, of all the fighting men of history, few are more impressive than the Greek hoplite of the fifth and fourth centuries BC and few can be as picturesque. With their crested helmets, huge round shields with variegated blazons, and their long spears, a body of such warriors must certainly have provided a tremendous spectacle, one calculated to impress any foe with the liveliest apprehension, for, apart from the matter of appearance, they were no mean fighting men, as they proved on not a few battlefields, both in Greece and in neighbouring countries. Throughout history, certain types of soldier have become strongly identified with the national character and qualities of their country of origin—the Old Guard Grenadier seems to epitomise Napoleonic France, the rifle-armed Kentuckian provides the spirit of the American Revolution and so on—but none really provide such an apotheosis of their age as does the hoplite of Classic Greece.

This possibly results in part from their being—in the early days at least—recruited from the citizenry of their own particular city, and thus they were rather more immediately identified with the spirit of their 'polis' than were the troops of other areas and periods. The reasons for the origin and development of the heavy infantryman we call the hoplite are fairly complex and combine the political, the economic and the topographical. It is sufficient to say, however, that the old Heroic Age of Greece—the great Homeric period—with its warfare involving chariot-borne kings—'heroes' if you like—which was traditionally at its height round about the time of the Siege of Troy, gave way to a much more democratic system as the city-states of Greece—Athens, Thebes and Corinth for example—developed economically and commercially and a middle class emerged. From this came the citizen soldiers who provided the great proportion of the defenders of each individual state. Now, because these soldiers were essentially part-timers and as their periods of training and drill

16

were fairly rare, there developed over the years (and it must be stressed that the hoplite did not appear overnight, but came into existence during a fairly lengthy period) a relatively simple tactical formation, to wit, the phalanx. In this, all available manpower—or at least the manpower able to provide the necessary armour and weaponry—was massed together in a tightly organised array in close order with a depth of several ranks, sometimes 6, sometimes 8, a formation which, although unwieldy and inappropriate for really complicated manoeuvres, was nevertheless a frightening thing to a prospective foe by reason of its sheer mass and solidity. As all the members of the phalanx were citizen soldiers of some substance, it was a matter of pride that they should provide themselves with the best possible armour and weapons, so that virtually all had the basic necessities of helmet, shield and spear which provided a nearly impenetrable front to the hoplite phalanx. Basically, the phalanx was a shock weapon, designed for a straightforward charge home, this being its inevitable tactic resulting from the lack of ability to manoeuvre.

Having, in fact, developed through the sixth century BC, it was really in the early part of the fifth that we have the beginning of what might be called the Age of the Hoplite, a period wherein warfare in Greece and the Eastern Mediterranean was dominated by this type of warrior. It was the great battles of Marathon and Plataea in the first quarter of the century which showed just what the hoplite was capable of in circumstances favourable to him, and the great enemy, Persia, could not prevail against the densely arrayed phalanx of heavily armed and well-armoured hoplites, bearing the 'hoplon'—whence they derived their name. It is at this point in the period of hoplite domination—which lasted something like the best part of two centuries—that we can commence our study in miniature of this fascinating fighting man, and we can do this with some advantage by first examining his arms and equipment as they might have been when the Athenian tribes marched out to engage the might of Persia at Marathon. *Fig. 2.1* shows such a hoplite—he is really a synthesis of probabilities, based on a variety of Greek vase paintings—and it would be appropriate first to consider the most eye-catching portion of his panoply, the hoplon or shield.

This was a massive affair, and when the hoplites were drawn up in close formation, each man's shield slightly overlapped that of his left hand neighbour, thus forming a continuous shield wall and no mistake. The weight of such a shield must have been something of a problem, and such an artifact in solid bronze, the favourite metal, would have been ponderous in the extreme. Thus, wood was used in its construction, sometimes faced with a thin sheet of bronze. A sheet of

Fig. 2.1   A Hoplite

bronze, backed by a stout layer of boiled leather—*cuir bouilli*—was also employed, this being a combination both light and extremely tough. There does not seem to be any suggestion, by the way, that the hoplon was itself used as a secondary weapon (as was the shield of the Roman legionary in later times) and in action it was moved into position in front of the body, held there by two attachments on the back (*Fig. 2.2* illustrates). Across the inside diameter of the 'dished' shield was the 'porpax'. The arm was inserted through this so that the bar was just in front of the elbow, while the hand grasped a leather or rope—the 'antilabe'—which in some cases was continued right round the inside of the hoplon, as in the illustration. Presumably, should one hand-hold break or be severed, another close at hand could be grasped in lieu.

Of course, as always throughout history, fighting men have found it advisable to identify themselves, and the Greek hoplite is no

Fig. 2.2   The 'hoplon

exception to this rule. All sorts of devices can be seen on the hoplon (Greek vase paintings show them frequently, sometimes in the greatest detail) and they varied—as we see in *Fig. 2.3*—from the highly elaborate to the severely functional. In many cases, with the

Fig. 2.3 'Hoplon' devices

hoplite supplying his own 'gear', the shield 'blazon' was a very personal one, but, as time progressed, certain city states followed the example of Sparta, the shields of whose hoplites all bore the single Greek letter, the lambda (*Λ*)—for Lacedaemon, as the Spartans called their state. Sicyon and Messenia were two such cities who adopted the initial for use on their shields. Another example was Thebes, whose shields were ornamented with a club, the device of Hercules. It seems that there were two methods of blazoning the shield. Sometimes the device, whatever it was, was simply painted upon its face, but there have been found examples where the device was cut from a thin sheet of bronze, this itself being affixed to the front surface of the shield. From various references it would appear that the bronze face of the hoplon was highly polished, and one occasionally sees representations of a kind of leather apron hanging below the shield, presumably to afford additional protection to the legs, but this practice does not seem to have been a common one.

Having thus briefly considered the piece of equipment which gave the hoplite his common name, it seems that the next most characteristic, and certainly the most eyecatching feature about him would be his helmet, and a most impressive object this generally is, although, as time passed, it became more functional and less flamboyant than it was originally. Now, this sort of headgear seems to have been influenced initially by Assyrian helmets, at least as far as the characteristic crest is concerned, one of the earliest examples (found at Argos) being quite definitely Assyrian in type, conical and with a curved crest elevated on a species of 'stilt' on the top of the helmet. From this derives the helmets seen in many classic Greek vase paintings and sculptures, but before long we find the crest attached directly to the crown of the helmet, which is itself an all-enclosing affair such as is worn by our hoplite in *Fig. 2.1*. This is the type known as the Corinthian, a remarkable product of the armourer's skill, being beaten out, it appears, from a single sheet of bronze, an art which disappeared for many centuries after the hoplite had passed into history. The Corinthian helmet must have been a dreadful thing to wear, and with his head encased in it the hoplite must have been virtually deaf and had his vision restricted to a narrow area directly to his front so that it is hardly surprising therefore that considerable changes were made to ensure that he could at least hear the commands of whatever leaders he had and that he would have a clearer view of what was going on about him. Various forms of helmets—for instance the 'Illyrian' and the 'Chalcidian'—were to be seen as early as the sixth century BC but we have to limit ourselves to noting a couple of variations possibly of more interest to the student. The first of these is the so-called 'pilos', which existed towards the end of the fifth century (we show it in *Fig. 2.4*) a much smaller and simpler thing than the Corinthian. It was initially, it seems, associated with Sparta but there is monumental evidence that it was worn by Athenians. It is discussed at some length by Professor J. K. Anderson (*Military Theory*

2.5   The Boeotian helmet   Fig. 2.4   The 'pilos'

and Practice in the Age of Xenophon, University of California Press, 1970) and his suggestion—which seems a most logical one—is that it was originally the helmet lining, first used by itself as a hat, then later copied in metal, and he gives examples of bronze piloi which have been found here and there in Greece. Indeed he mentions that in one of the plays of Aristophanes a cavalryman is mocked for using his pilos as a cooking receptacle (reminiscent of the old 'tin hat' or the 'battle bowler') One such bronze pilos helmet, found at Piraeus, is illustrated in Peter Green's *Alexander the Great* and there can be no doubt that it must have been a far more convenient article than the Corinthian or any of its derivatives.

The other type—really too distinctive to be called a variation of the Corinthian—is that typical of the district of Boeotia in north-eastern Greece. To all intents and purposes it is a wide-brimmed metal hat with the brim bent down at the sides, while the portion in front is untouched, to allow the wearer better vision (*Fig. 2.5* illustrates the Boeotian helmet). During the fifth century BC Boeotian hoplites are distinguished by their wearing of this sort of helmet and a century and more later the Greek military writer Xenophon recommended it for cavalry use. It was in fairly general use for both horse and foot during the campaigns of Alexander the Great (one was found in the River Tigris and there are representations of more than one of Alexander's 'paladins' wearing the Boeotian helmet on the 'Alexander Sarcophagus' at Sidon). Despite all these changes it would appear that the Greek hoplite—like many military men—was a conservative individual and even at the end of his period of domination—say the end of the fourth century—he still clung to the trappings of former days, adapted and altered though they might have been. As for the helmet, there remains but one point of interest. As far as I know all crests on Corinthian helmets run fore-and-aft, i.e. from front to rear. There does seem to be one exception, seen on an early fifth century bronze figure of a warrior, considered, by reason of the long tresses of braided hair to represent a Spartan (this being a Spartan characteristic—see Thermopylae, where the Persian scouts were surprised to see their Lacedaemonian opponents combing and braiding their

Fig. 2.6
The Corinthian helmet

hair prior to being attacked). *Fig. 2.6* shows this warrior wearing the Corinthian helmet ornamented with a very distinctive transverse horse-hair crest. This is a comparatively rare phenomenon, but similar examples are referred to by H. R. Robinson in *The Armour of Imperial Rome*, but the great majority of Greek helmet crests are of the predictable fore-and-aft type.

So far so good, then, and the next article we consider in our study of hoplite equipment—only briefly, alas—is the cuirass or corselet. During the early period, the seventh and sixth centuries, this was a very massive article indeed—the *bell cuirass*—consisting of very heavy bronze plates for breast and back, flared at the lower edge—thus the name. It must have been more than a little exhausting to wear for any length of time, and by the time of the Persian invasions—early fifth century—it had been replaced by a much more convenient type, the 'composite corselet' with 'epomides', such as is worn by our hoplite in *Fig. 1.1*. Such a defence would be both convenient and efficacious, and must have seen much service throughout the Eastern Mediterranean region. Still later, there came into use the 'muscle' type of cuirass, consisting of front and back pieces moulded to reproduce the body muscles. I feel that this type, necessitating fairly accurate fitting to the individual wearer, was not too widely used, being confined perhaps to the wealthier of the hoplite class. It could have been, by the way, either leather or metal, or possibly the former was painted or gilded to represent metal. In point of fact, as the years passed the tendency was for hoplite equipment to become lighter, and during the fourth century BC the cuirass partly at least gave way to the 'spolas', but this is really taking us a trifle out of our classic hoplite period and we shall say no more about it here.

Thus we come to the final article of armour—the greaves to protect the lower legs. They were of metal, shaped to fit the limb, and in manufacture beaten out until they were thin enough to be attached by the 'spring', being fastened to the leg rather like bicycle clips. To prevent chafing they were lined and in surviving examples one can see holes round the edges whereby the lining was attached. In its own way, the greave at this time was nearly as much an armourer's triumph as was the Corinthian helmet.

So, at this point we can look at our hoplite from a defensive point of view and this may be illustrated by *Fig. 2.7*, which shows a few hoplites drawn up in battle order. I think it provides some idea of just how invulnerable a hoplite phalanx must have appeared, only the very minimum of human body being exposed. Truly the approach of such a mass of men must have had a more than daunting effect on whosoever was braced to oppose it, especially when one adds the serried ranks of spearheads projecting from above the shields. These

Fig. 2.7   The Hoplite phalanx

spears were, incidentally, the principal weapon of the hoplite, and they were usually iron headed, some 8 or 9 ft in length, and often shod with bronze. Although one sees representations of individual combats with the spear held underarm, in an advance *en masse* it was held overarm, and with the closeness of the ranks 2 and possibly 3 rows of spear points projected over the front shields. At one time an additional spear or possibly a javelin seems to have been carried, but this was abandoned and the single, heavy thrusting spear became the primary weapon.

Apart from the spear, which he naturally favoured, the hoplite was furnished with secondary armament, usually the sword, this itself having several interesting features. Originally, it was straight, with a somewhat leaf-shaped blade, and a cross-piece on the hilt, the whole affair being about 2 ft in length, or possibly a little less. Many sixth century vase paintings show it as being worn from a cross belt and carried fairly high up on the lefthand side of the wearer. Even allowing for artist's licence, some of the hilts are pretty elaborate, ornamented with round or square bosses (see *Fig. 2.8*). However, this weapon seems to have been largely supplanted by another, which seems to have gained popularity during the fifth century. It was a

Fig. 2.8a. The 2ft sword

Fig. 2.8b. A 'Kopis' or 'Machaira'

slightly curved, slashing weapon, bearing a considerable resemblance to the Gurkha *kukri* (see *Fig. 2.8b*). It had a very heavy back to the cutting edge, which was on the convex side, and was characterised by a distinctive hilt with a half-guard (this seems the only way to describe it) ending in what appears to be the head of a bird. This must have been a most formidable weapon, its weight, as in the case of the *kukri*, imparting tremendous force to a downward slash. The name of this sword presents something of a puzzle, and scholarly opinion is divided. It is believed by some to come within the term 'kopis', but it could also be called a 'machaira', a type which was later recommended for cavalry use by Xenophon. This, by the way, is not the only occasion in ancient history when it is difficult to determine precisely what some word really means. Alas, no one thought fit at the time to illustrate his use of a particular word, or to caption a vase painting. My own feeling—and I am subject to instant correction—is that it may be the machaira, and that it was with such a sword that Black Cleitus struck off the arm of the Persian Spithridates when he was attacking Alexander at The Granicus.

At any rate, there you have our hoplite in his full and undeniably impressive panoply, individually in *Fig. 2.1* and as a small section of a fighting unit in *Fig. 2.7*. The question at once arises, naturally, as to just how effective he was. In suitable circumstances, especially on terrain which favoured him, there is no doubt but that he was a very tough customer, and during the years in which he dominated warfare there are many occasions when he amply demonstrated his prowess. In considering this, it must not be forgotten that in many cases he was pretty well an amateur—a part time soldier. From this one must naturally exclude the professionals—the mercenaries who, in increasing numbers, were to be found in the most surprising places during the fifth and fourth centuries, and, of course, the Spartans.

The latter defended the only city-state where the military art and its discipline were constantly practised and were indeed something of a religion. Considerable evidence has come down to us of the not un-complicated manoeuvres which could be carried out by Spartan infantry, and it would seem probable that groups of mercenary hop-lites fighting together for a lengthy period would also develop similar skills. Certainly the 'Ten Thousand' of Xenophon's *Anabasis* must have been such and their incredible retreat from the depths of Persia to the sea must stand as by far the greatest exploit with which the Greek hoplite can be credited. Even so, one must not overlook the fact that, as shock infantry, they had one primary function and this was to close with the enemy as soon as possible and overthrow him by weight and the momentum of the charge.

The business of the charge itself is an interesting one, for contrary to what might be thought, it must have been carried out at no little speed, at least a fast trot. Hoplites were eminently capable of such movement for at Marathon, according to Herodotus, the Athenian phalanx 'advanced at a run' towards the enemy, who were something like a mile away—this being largely for the purpose of passing through the 'beaten zone' where archery was at its most deadly. An-other instance comes from Xenophon, who relates that on his un-fortunate—for him at least—march to the Battle of Cunaxa, Cyrus was asked by the Queen of Cilicia to demonstrate the abilities of his mercenary Greek hoplites for her benefit. The phalanx accordingly paraded in all its panoply of bronze and scarlet and when the signal was given, the entire mass of 10,000 hoplites advanced, as it happened, either by accident or design towards the spot where the Queen sat in a carriage. The troops accelerated until they were actually running to-wards the royal equipage—'. . . the natives were terrified, the Queen of Cilicia fled . . .' and the Greeks returned to camp highly diverted at the incident.

We have said that the hoplite phalanx ordinarily formed up 6 or 8 ranks deep, although this was reduced at times, and for long this was the norm, particularly during the period of Spartan domination. This supremacy was, however, destroyed by the hoplites of another Greek state, Thebes, whose practice it was to draw the phalanx up in a much deeper formation than was customary with other city states. With a 25-man deep formation they defeated the opposed Athenians at Delium in 424 BC. This idea was exploited by their general, the celebrated Epaminondas, and at the Battle of Leuctra (317 BC) he massed a column of hoplites no less than 50 deep on his left, at the same time 'refusing' his right. This column, whose charge was com-pared by Xenophon to that of a trireme at sea, shattered the Spartans, despite their being 12 deep. This feat was repeated at Mantinea (362

BC) and from this defeat Sparta never recovered (Epaminondas, sad to say, was killed at Mantinea). However, Macedon soon took over the controlling voice in Greek affairs.

The success of this extra-deep formation is interesting and provokes some thought. We can picture two opposing hoplite phalanxes in a mutual charge, coming together rapidly and meeting with an almighty crash as shield clashed with shield and the long spears thrust at what little of an enemy was exposed. Now, in a hoplite versus hoplite fight there would initially be few casualties, but here and there one or two would fall and finally one side would turn and flee, at which point the heaviest losses would be suffered as unshielded backs would be turned to the enemy. Such a fight might be likened to a rugby football scrum, with two sides heaving to and fro until one gives way. Now, if instead of only three rows of forwards, one side had ten rows, obviously, bound together, the weight and push of the ten-row scrum would prevail, without a doubt, but in a hoplite phalanx, there was no such binding, so how exactly did the 50 ranks prevail over the 12? Writers speak glibly of the 'weight', but this cannot be so if all ranks are not pushing, and I cannot see 49 men all pushing against the chap in front with their shields—highly uncomfortable and no mistake. So was the effect a morale one, or did the existence of the additional ranks provide more replacements for casualties up front—or what? It is an interesting question for often one takes for granted in military history statements which may need much explanation. Or at least I feel so.

In any case, this was the classic Greek hoplite and a prestigious fighting man he was; of that there is no doubt.

III THE BATTLE OF MARATHON 490 BC

Greeks v Persians

THE ORIGINS of the Marathon campaign and the dramatic battle in which it culminated are to be found during the previous century and they derive quite simply from the expansion—in particular the westwards march—of the Persian Empire.

Founded by Cyrus the Great, whose policies had been amply fulfilled by his equally aggressive successors, the Empire was in fact the greatest the Western World had yet seen, at its maximum extending from the borders of India to the Mediterranean and from the Black Sea to Egypt, and—here we come to the very nub of the matter with which we are concerned—including the numerous Greek colonies then distributed thickly along the coast of Asia Minor. The Persians were high-handed and autocratic in their dealings with those they considered their inferiors and their treatment of the Greek colonists was far from being to the latters' liking. Increasing discontent culminated in open rebellion in 499 BC, when the people of Ionia, as the Greek fringe was called, made a concerted effort to free themselves from the harsh rule of the Persian overlords. They marched upon, and seized the city of Sardis, but were unable to occupy the great Acropolis there, and, possibly accidentally, at the same time the city was burned to the ground.

It was really this incident which was the spark to set alight the tinder, it being—in his eyes—a grievous insult and affront to his dignity which King Darius of Persia could not possibly overlook. However, it was some time before the Persian army—never the speediest of campaigning forces—was able to cope with the revolt, but it was finally crushed in 495 BC, when Darius' ships destroyed the Ionian fleet in a sea battle off the Gulf of Miletus. Following this success the Persian ships instituted a species of blockade of mainland Greece—whence it was considered that help had come to the Ionians in their struggle—the aim being to stop the grain ships coming from

28

the Black Sea, a trade vital to every city in Greece. This was a long term policy, however, and it was but preliminary to the main action.

In 492 BC a large fleet and army were assembled by Darius, the command confided to Mardonius, his son-in-law, and sent across the Dardanelles. Fortune was against the enterprise, though, the army itself being roughly handled on its entry into Thrace, and at the same time, the fleet was largely destroyed by a storm off the Athos peninsula. Mardonius was glad to make his way back to Persia.

This expedition was really only in the nature of a curtain-raiser, for at once Darius, with renewed determination, set about assembling more ships and troops, at the same time indulging in a sort of propaganda war against the northern Greek states by sending envoys to them to demand 'earth and water', this being a token of their accepting vassalage. Many gave way to the threat and soon the greater part of Thessaly had submitted to Persian demands. The groundwork of his campaign thus being laid, Darius believed that the time was ripe and, considering that the material he had collected in the way of men and ships was sufficient for his purpose, he gave the command of these powers to his nephew Artaphernes and a Median noble named Datis. Under their direction the fleet set sail with orders to reduce Athens and Eretria (a city on the island of Euboea) to slavery, these two states having been singled out for special attention because they were believed to have been the principal furnishers of aid to the rebels during the Ionian revolt. A landing was made on Euboea by the Persian army—a force which had received several accessions of strength *en route*, the whole now amounting to some 25,000 men in about 600 ships. After a brief space of time, Eretria was betrayed from within and in accordance with the will of Darius, all the temples in the city were destroyed as a mark of his displeasure. Following this initial success the Persians sailed for mainland Greece, their destination being Marathon, a coastal place something like 25 miles from Athens itself. This would probably be about the beginning of September, 490 BC.

The choice of the landing ground was far from fortuitous. Among the Persian host was one Hippias, an Athenian, but an exile from his native city. He was in touch with dissidents therein, those making up what in latter day parlance would have been called a 'fifth column', and his hope was that, with the help of these traitorous elements, and backed by the Persians, he would be reinstated to his one time position of power in the 'polis' of his origin. It was he indeed who was responsible for the choice of beachhead, as it led on to a long stretch of flat ground highly suitable for cavalry operations in addition to having an excellent beach, upon which the Persians could draw up their vessels. The approaches to the position were limited to two narrow valleys through the surrounding hills, and all in all it seemed to the Persian leaders, Datis and Artaphernes, that they were in an excellent situation, politically as well as militarily.

Understandably, at Athens, which had been apprised of the landing by a system of alarm beacons, there had initially been considerable confusion. It must be remembered that at this time Persia bulked largely and fearfully in the Greek mind, to many Athenians seeming to be quite invincible; and due to the persuasion of genuine pacifists as well as of the fifth column already spoken of there was considerable pressure upon the city fathers for an outright surrender. Nevertheless, it was decided to send for help and a runner, by name Pheidipides, was sent off with an appeal to Sparta, where he duly arrived, having covered the 140 miles in a couple of days or so. The Spartans, a very religious people, were apparently eager to help, but claimed that no action could be taken until after the conclusion of an imminent religious festival. It may be unjust, but it was not the last time that the Lacedaemonians—or Spartans, as we more generally know them— were to postpone a vital decision because of a convenient religious festival.

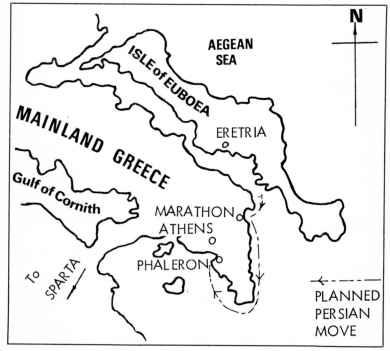

Fig. 3.1   Persian Strategy

In the meantime, however, urged largely to the decision by one of their leaders, Miltiades, the Athenians decided to take the bull by the horns and meet the threat much more than just half-way. Accordingly, 10,000 bronze-panoplied hoplites—heavily armed and armoured infantry—marched out of the city *en route* for Marathon, under the overall command of the War Archon or Polemarch, Callimachus of Aphidna. With this formidable body went numbers of slaves and lightly armed people to afford some help, although these naturally had little or no fighting value. Marching with some expedition by the coastal route, the Athenian force arrived at the western end of the Marathon plain and took up the position shown on *Fig. 3.2*. At this point the general position, including the whereabouts of the Persians. may also be seen on the same map. The situation appeared to favour the invaders, for the plain was ideal for manoeuvre by cavalry—an arm completely lacking in the Greek army—being interrupted only by occasional scattered trees and clumps of bushes. To the north-east was a large marsh and the whole plain was surrounded by wooded hills rising to mountains. The position of the River Charadra is shown, but at the time of the battle it must have been almost completely dried up and it certainly does not figure as an obstacle in any account of the battle which was to follow. In their camp the Athenians were joined by about a thousand hoplites from the allied city of Plataea, bringing the strength of the Greek army to about 11,000 men. Although the march of the Greeks was a bold one it was in actual fact fraught with peril as it seems to have fallen in almost perfectly with Persian strategy. This, in short, was to entice the main Athenian field army away from the city, 'pin' it with a covering force to enable the main body to re-embark and set sail directly to a landing as near as possible to Athens—this being Phaleron—then march on the city, whose gates were to be opened by the traitors within (*Fig. 3.1* shows the elements of this strategy). However, not for the first nor the last time, men's best laid schemes failed to come to their desired fruition, foundering on the bold approach and the effective action of the Athenian army.

At this point some discussion may be in order concerning the strength of the two armies. We have already noted that, with the advent of the Plataeans, the Greeks totalled some 11,000 men, all armoured and shielded hoplites, bearing the 9-ft spear plus a sword, wearing helmet and either leather or metal body armour, and equipped with the great round shield—the hoplon—from whence came their name. There was possibly some rather meagre support in the shape of the camp followers who accompanied them. It will be no great surprise to find considerable variations in the estimate of historians concerning the Persian numbers, these ranging from a .

total of about 15,000 to nearly 25,000, and giving the cavalry component as anything from 1,000 to 5,000 men. The former figure is accepted by Fuller (*Decisive Battles of the Western World*, Vol. 1, page 20) and the latter by K. P. Kontorlis (*The Battle of Marathon*, 1973), while Peter Green (*The Year of Salamis*) is silent upon the subject. However, in a brief but penetrating article in *Slingshot* (Journal of the Society of Ancients, September issue, 1973), R. B. Nelson demonstrates most convincingly, by deduction based upon the number of transports in the Persian fleet and their horse-carrying capacity, that the probable maximum was indeed not more than some 800.

All in all, considering the various sources and estimations it seems reasonable to assume that the Persian army numbered some 15,000 infantry, and simply for the convenience of using round figures, about 1,000 horse. Generally, neither the Persian cavalry nor infantry could match the Greeks in either weaponry or armour. The bow was almost a universal weapon, eked out by a few javelins, while only a relative few wore any sort of armour. Any shields the Persians carried were fragile in comparison with the hoplon of the Greeks. In fighting ability, however, the best of the Persians, as well as the Medes, were of high quality and, in courage, were in no way inferior to their enemies.

Thus, at this point of the action, this Persian army was posted in front of the Great Marsh, with the fleet lying as shown by the Schoinia beach.

It was, in fact, something of a stalemate situation. The Greeks were in a most advantageous position defensively, drawn up on rough, sloping ground where the Persian cavalry could not possibly get to them without being hopelessly disorganised. The Persians, in fact, with this cavalry and their numerous archers, would have welcomed a Greek advance across the open plain to make a direct contact. Thus, both sides remained in this position, watching each other, for over a week, when the capture of Eretria became known to the Athenians. Matters had indubitably reached crisis point and the sight of the considerable Persian fleet brought home to the Greeks the fact that an immediate descent upon Athens by sea was a very lively possibility and that something would have to be done about it and that right speedily.

The news concerning the fall of Eretria may have coincided with the arrival of information that Artaphernes was about to embark his army, probably starting with the cavalry, which would be needed for a quick rush for Athens when a landing had been effected at Phaleron. Certainly cavalry was not present in the early stages of the battle which followed, although Kontorlis claims that it did participate at

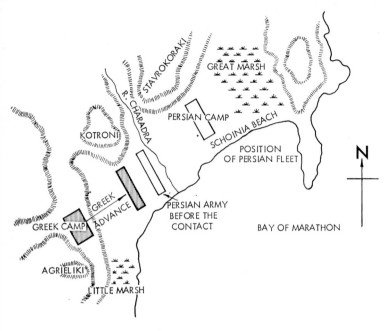

Fig. 3.2   The elements of the battle

its conclusion, covering the embarkation, and he points to a relief on the Brescia Sarcophagus which shows mounted Persians as well as infantry fighting by the ships on the beach. It may be just possible, at the beginning of the battle, that part of the cavalry had, in fact, embarked or was about to do so, and that some were able to take part in the final struggle before the fleet weighed anchor and made for the open sea.

One other problem remains. This concerns the position adopted by the armies just before the clash. The older proposition was that the Persian line was parallel to the beach and that the Greeks came towards them directly from the north. This, on the whole, seems unlikely. It appears hardly feasible that the Persians would adopt a position which more or less presented a flank towards the first Greek position, and which would have allowed a 'side step' by an enemy coming from the north to enable them to get between the Persians and the beachhead with its vulnerable shipping. Peter Green (*Year of Salamis*) describes the Persian line as awaiting the Greek attack on a line 'reaching from Mount Kotroni to the shore', with archers in front and some cavalry on the wings, although he does imply that these must have been very few in number, if indeed there were any at all. R. B. Nelson, in the *Slingshot* article referred to above, agrees with this, and Kontorlis (*The Battle of Marathon*) confirms this view by pointing out that, when it was discovered in recent years, the burial mound of the Plataeans, who formed the left wing, showed that the Greek battle line was at right angles to the shore. Incidentally, the position of the Persians from Mount Kotroni to the sea indicates that they had advanced quite considerably from their beach-head.

Probably early in the morning of the 12th August the Greeks made their decision to attack. This had been strongly urged by Miltiades, and when the Polemarch Callimachus gave his vote in support of him, the Greek army moved forwards from the wooded position they had occupied for the previous week or more in two columns which, when they reached the point of deployment, wheeled outwards to enable the head of each column to form the flank of the line. The Plataeans under Aemnistus were on the extreme left and, stretching towards the right were the ten Athenian tribal regiments, the Leontids and the Antiochids in the centre. Miltiades, as 'general of the day', had realised that his army would probably be outflanked by the greater length of the Persian array and that traditionally the best Persian troops would be in the centre. He therefore decided to concentrate on their less effective flanks and allow the strong centre to be 'contained'.

To this end he reduced his centre to a depth of only four ranks, while maintaining his wings at eight, thus simultaneously extending his front to match that of the enemy and ensuring that his strong flanks would be able to deal expeditiously with their weaker Persian opposite numbers. Thus, with Callimachus on the right, Miltiades himself with his own regiment, and with such famous names as Themistocles and Aristeides figuring in the centre regiments, the Greeks began to move against the Persian masses. Ordinarily at this stage the Greeks would have broken into their battle hymn or 'paean', but on this occasion it seems that they advanced silently, concentrating on the business in hand.

The danger, of course, was the arrow storm which the massed Persian bowmen would unleash as soon as the hoplites came into range, this being about 200 yards, and it seems that, under orders, they accelerated to a 'double' as they passed through the 'beaten zone', to shorten the period during which they would be exposed to this heavy fire. Their armour and speed reduced casualties considerably in the time occupied in closing with the enemy, and the ensuing combat was violent in the extreme.

As indeed was to be expected, the high quality Persian troops in the centre forced back the four-deep line facing them, some accounts actually suggesting there was a break, but this does not seem too likely, and it seems more probable that the centre of the Greeks simply recoiled before the intense Persian pressure. However, on each side the powerful Greek wings drove forward inexorably against their lesser opponents and before long both Persian wings were in full flight with the Greeks after them. At this point, however, due to a very high degree of discipline, the Greeks broke off their pursuit—an extremely difficult thing to do in the heat of action—quickly re-organised, and turned inwards to take the hitherto successful Persian centre on the flanks, bringing it to a sudden halt. This onslaught was too much for the Asiatics, who after a brief resistance, broke and fled. The pursuit was closely maintained, dreadful losses being inflicted on the fleeing Persians, many of whom were lost in trying to

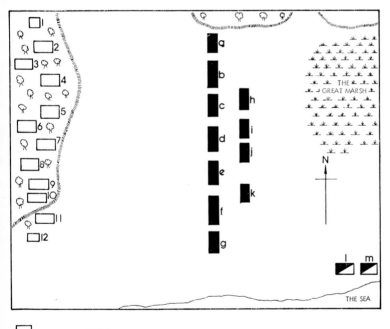

GREEKS
PERSIAN INFANTRY
PERSIAN CAVALRY

Fig. 3.3 Initial dispositions

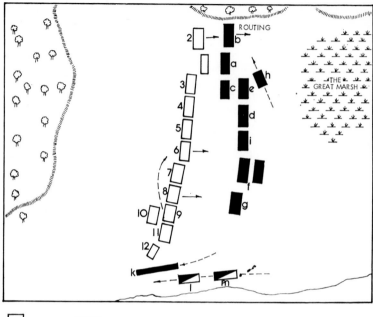

GREEKS

PERSIAN INFANTRY

. PERSIAN CAVALRY

Fig. 3.4   End of Period Eight

make their way through the Great Marsh. Others were cut down on
the beaches as they attempted to gain the safety of their ships. The
Polemarch Callimachus was unfortunately killed in the fighting, and
Artaphernes was able to get a part of his surviving troops on board
before the ships stood out to sea.

At a cost of less than 200 Athenians, and with over 6,000 dead and
many prisoners among the Persians, the Greeks had decisively flung
back the first attempt of the Great King to impose his rule upon their
country.

The wargame re-fight

No difficulty need be experienced in arranging the wargames table
for this battle. The area is well defined, needing little in the way of

accessories to reproduce the salient features of the terrain, while in shape the actual battlefield is more or less the rectangular area provided by the 9-ft × 7-ft table we used. In fact, by far the simplest procedure is to illustrate just how the table was set out, and *Fig. 3.3*, it is hoped, shows this reasonably well. As we contemplate it then, we see on the left the low, wooded hill where the Greeks had their encampment, while to the north is a suggestion of rugged foothills leading to the much more precipitous mountains behind. These northern hills were considered too steep for the movement of either infantry or cavalry, the plain providing no such impediment. The Great Marsh was, of course, a highly important feature, and its extent can be seen to be considerable. Such an area can be represented in various ways, that in use being simply a sheet of cartridge paper painted—in an endeavour to represent boggy ground—in somewhat vertiginous shades of green and blue, and fastened to the table with drawing pins. In actual fact, the southern edge of the table could have represented the coastline, but for the sake of appearance, the line of the beach was included—not a difficult process, a length of string being pinned into place as shown. Possibly the course of the River Charadra might have been included—enough Bellona stream sections were available—but as we have already indicated that it was sufficiently dried up to provide no obstacle to troop movement— indeed it is not mentioned in any battle account—it was omitted. One final and important point concerning the terrain should be noted, this being that the Great Marsh was passable but was a disorganisation factor in that infantry movement within its limits was reduced by half while that of cavalry, had it been appropriate, would have suffered a reduction of three-quarters. Such disorganisation would naturally have affected the fighting ability of any troops concerned but, as events were to prove, this did not have any bearing on the action.

There was no problem in assembling the figures to make up the respective armies, both Greeks and Persians being available in some quantity. It was then simply a matter of determining the number of units and the area they occupied and then in apportioning the figures to fill up the relevant spaces. First let us take the Greeks, placed at the outset of the battle as they are shown in *Fig. 3.3*, deployed along the wooded slope where they had been encamped for about a week. We know that the Athenians were organised militarily in 10 tribes, and having the assistance of the Plataeans, there are actually 11 regimental units to account for—the numerical key will serve to identify them on the maps of the battle at its various stages.

The Greek Army (hoplites)

2	The Plataeans	20
3	Akamantis Regiment	30
4	Oineis Regiment	30
5	Kekropis Regiment	30
6	Hippothonteis Regiment	40
7	Antiochis Regiment	30
8	Leontis Regiment	30
9	Pandionis Regiment	30
10	Aegeis Regiment	20
11	Erechtheis Regiment	20
	Total	280

In connection with this total—280 figures—two points have to be mentioned. First, for reasons of space and convenience, only 9 of the 10 Athenian tribal regiments are included, a forgivable omission, it is hoped. This resulted in there being 10 regiments of hoplites, these being all of the 'heavy' variety, bronze-armoured, equipped with long spear, helmet and shield. The absence of unit 1 in the list above will not have escaped the reader's attention. The explanation is this. It will be recalled from our background account of the battle that servants and camp followers in considerable numbers had accompanied the Athenian hoplites on the march to Marathon. It seems hardly likely that all of them would have been weaponless and it was felt that out of the number some would have been prepared to strike a blow for their masters or for Athens (after all, there is record of the Spartan slaves—the despised 'helots'—actually fighting in support of their masters on more than one occasion). Thus it seemed logical to add a couple of small units of these people to the Greek roster, it being decided that each should consist of ten javelinmen. To prevent their being too potent a force, they were graded as 'unwilling levies', not up to much indeed but nevertheless providing a minimal screen for the flanks of the heavy array of hoplites. Hence they take their places on the map—numbered as units 1 and 12, named the Messenion and Antelope Javelins respectively, and counting 10 figures each. They brought overall Greek strength to 300 figures.

Having established the composition of the Greek Army and drawn it up as shown in *Fig. 3.3*—this indeed being the position as chosen by the players—we come to the Persians. Numbers here are easy to determine as they are based on a figure we arrive at in relation to the strength of the Greeks. We have already accepted 15,000 for the

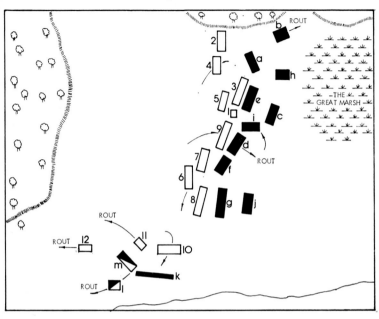

Fig. 3.5 End of Period Fifteen

strength of the Persian infantry; therefore, knowing the Greek hoplites to number 11,000, a simple calculation based on the Greek: Persian ratio of 11:15 reveals that, with 280 representing the Greek heavy infantry, there must be 400 Persian infantry figures. We have therefore the following impressive array taking up its battle positions.

The Persian Army

A	Ecbatana Infantry (light javelin-men)	30
B	Shiraz Infantry (bow, spear and shield)	40
C	Arachusians (light archers)	30
D	Phrygian Infantry (shield, spear)	40
E	Parsa Infantry (bow, spear and shield)	40
F	Median Infantry (bow, spear and shield)	40
G	Mudrayan Infantry (bow, spear and shield)	40
H	Lebedos Infantry	40
I	Samos Infantry	40
J	Lesbos Infantry (H, I, J: Ionian hoplites, leather armour, spear and shield)	40
K	Kirmanshah Light Infantry (bow, spear and shield)	20
	Total	400

This then is the Persian Army as it was deployed at the beginning of the battle, left flank on the beach and right flank on the high ground to the north. The Ionian levies form the bulk of the second line and they, like the Greek javelinmen in the other army, were graded as 'unwilling', of no great fighting value. Their real value was hardly up to their numbers and appearance. As an example, a single 'push back' in mêlée would be enough to make them break and run, in contrast to the Athenian hoplites who required three similar reverses before behaving in this reprehensible fashion. There is, by the way, no real historical justification for their nomenclature. Such details are not on record and the names were chosen more or less at random from Greek cities in Ionia. It might well be that some levies may indeed have come from the places with whose names I have been so free. Generally the Persian native infantry ranged from the excellent to the mediocre and it goes without saying that their missile power was formidable, no less than 210 of the figures being armed with the bow. Without a doubt it was patent that any direct attack would bring down a hail of arrows upon the assailants.

Now we come to something of a knotty problem. It concerns the Persian cavalry, which we have suggested was, in round figures, no more than 1,000 men. Already we have discussed the question as to whether or not any cavalry actually took part in the battle, but what number did, if any, remains a mystery. The problem therefore was whether Persian cavalry should be present, and if so, where they were to be stationed. This caused much discussion and it was at length agreed to include 2 units of light cavalry, each of 10 men, which would in actual numbers equate fairly closely with the original number. These two—armed with bow and short spear—were posted on the beach some distance to the Persian rear, as though they were about to embark. It is confessed that this might have been something of a mistake as their presence meant that the original plan of the Greeks was abortive and they fought not only the Persian infantry but the cavalry as well. In any event, for better or worse, there was added to the Persian army list, the following:

| L | Sogdian Light Cavalry | 10 |
| M | Dahae Light Cavalry | 10 |

Their position can be seen on *Fig. 3.3*, and their addition brought the overall Persian strength to 420 figures.

So, with the armies in position we are nearly ready to proceed, only one point remaining to be noticed. It will be recalled that the Greeks were ordered to advance 'at the double' as they crossed the 'beaten zone' to reduce the effect of the arrow fire. It seemed proper

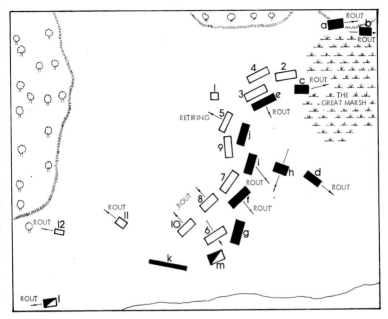

Fig. 3.6   Final Positions

that this should be duplicated and the Greek commanders were privately informed that their infantry—all 'heavy', with a move of 4 in. per period and a charge of 6 in.—would be allowed three moves as medium infantry and a charge move likewise, that is, 6 in. and 8 in. respectively, while at the same time retaining their defensive heavy infantry attributes. This was considered as adding to the realism of the proceedings and of course it was not immediately communicated to the Persian hierarchy, who took cognizance of it only when in the fullness of time the heavy hoplites made their first move as 'mediums'. (It may be of interest, from the practical wargaming point of view, to note that since the battle was fought, the wargame rules employed have been modified, and both medium and heavy hoplites enjoy similar mobility, that is, ordinary and charge moves of 6 in. and 8 in. respectively.)

At this time it may be considered appropriate to identify the commanders. The choice of role for each of the participating wargamers was determined simply by lot. The names of the Greek and Persian generals were written upon slips of paper, these being screwed up and deposited in a receptacle, whence one was drawn by each player in turn. Once identities had been decided, the generals decided among themselves which section of their armies would be under their immediate command. Upon the Greek side were 'Miltiades', whose mantle was donned by Mark Smith, 'Themistocles' was David Matthews, and finally 'Aemnistus' was Andrew Green, who naturally took command of the Greek left, which included his Plataeans. The scrap of paper bearing the name of the Polemarch Callimachus remained unclaimed. This was fortunate in a way because there were three available slips bearing Persian names, so that three players were most fortuitously ranged upon each side, the Persians being 'Datis' in the person of Alan Angus, who took over the right wing, 'Artaphernes' was Derek Casey, commanding the left, and—to make up a triumvirate of players—we included the renegade Greek 'Hippias', who as Peter Sheppard, commanded the second line of infantry, mainly the Ionians.

Both sets of generals were provided with brief 'sitreps', reading as follows:

Greeks: For seven days you have been observing the Persian Army from your camp, but because of the presence of enemy cavalry, you have not advanced. His mounted troops now appear to be embarking and you will advance, attack and drive them into the sea, to prevent their making a sea-borne attack on Athens. The enemy is a mixture of regular and barbarian troops, while yours are all regular, with the exception of the javelinmen.

Persians: For a week you have daily offered battle to the Greeks, but they have not left their position on the heights. It has therefore been decided to embark the army to make a descent upon Athens. The cavalry is about to board and the infantry will remain in position until it has done so. Some activity has been observed in the Greek camp.

With the scene thus set, the players proved eager to get on with things, and the battle began, as was to be expected, with a general Greek advance, not in the historical twin columns, but in a single formidable line, with the left flanking Messenian javelinmen moving diagonally across the hoplite front to provide a screen for the left wing. On the other wing the Antelope javelinmen also moved ahead. Nor did the Persian generals delay, having obviously come to the conclusion that the sooner they could bring their massed archery to bear on the Greeks the better. This they did as early as the second period, at long range it is true, but effectively enough to cause slight casualties to two of the Athenian regiments, one of which, the Antiochis, was so incensed that when a reaction test was taken, it went into 'uncontrolled advance'. The figure of 'Themistocles', however, was close at hand and was able to teach the over eager unit and give it fresh orders at the end of the mandatory two periods of uncontrolled advance (perhaps incorrectly the general figures were mounted, but it did give them greater freedom of movement).

Meantime, events of considerable importance were taking place on the Persian left. The Kirmanshah light infantry came forward to the edge of the sea and began to move along it in single file with the aim of getting on the flank of the Greeks, who were steadily advancing, while at the same time both Persian cavalry units were coming forward rapidly (16-in. moves, of course). It was possibly a trifle unfortunate that, at the time of the battle, 'Miltiades' was the least experienced of the players and that, in 'Artaphernes', he was faced by a veteran of many an ancient battle, as well as being a devious and cunning wargamer. 'Themistocles', who might have advised his colleague, was already fully occupied with the regiment out of control; 'Miltiades' was left to his own devices and the Persian outflanking move was allowed to proceed. On the other wing the Ecbatana javelins had been brought forward to engage the Messenians who, despite their poor fighting quality and their precarious morale, held out admirably, notwithstanding severe losses, By this time the Greeks, under arrow fire all along the line, were now advancing 'at the double', somewhat to the surprise of the Persians, who immediately began to pull back, hoping to prolong the shooting time, but the Shiraz regiment was slow and was charged by the Plataeans. After a period wherein they just held their own, they broke and fled. Their place was speedily filled by the Lebedos hoplites who had been moved up from the second line by the watchful 'Hippias'. In the Greek centre the Antiochis regiment, under orders again, had joined in the general advance, but the archery of Kirmanshah was telling on the Antelope javelins on the right, and the two cavalry regiments were about to pass the end of the line (*Fig. 3.4* shows the situation at the end of Period Eight).

Although the Messenian javelinmen had been reduced to only 2 figures out of the original 10, their morale, when tested, was *mirabile dictu*, still favourable, and they were indeed able to fling themselves upon the flank of the Phrygians now fighting Pandionis and having thus (see the rules) disorganised them, were largely instrumental in causing them to break. The reaction tests which followed this rout all along the affected portion of the Persian line were favourable to the Asiatics, although their entire right flank was under severe pressure and much of it was retiring, largely as a result of the destruction of Shiraz and the Phyrgians. There was far less satisfaction on the Greek right, however. The Sogdian and Dahae light cavalry, having been allowed to circumnavigate the Athenian right, were now harrassing the Erechtheis regiment and the Antelope javelinmen, who were also suffering from the fire of Kirmanshah, now squarely upon their flank. Indeed it was not long before the javelins broke and ran, leaving Erechtheis to cope with virtually an all-round attack.

The fight put up by the Persian centre was something of a surprise to the Greeks, the Parsa infantry, despite the Greeks' superiority in arms and armour, actually pushing back its opponents. Akamantis, being ably seconded by the Samos hoplites, just fed into the line by 'Hippias', held on stoutly against the Pandionis regiment. On the Persian right, though, the enemy weight was telling, and the line in this sector was almost at collapsing point, although in the confused fighting near the beach, on the other hand, losses had been heavy on both sides. In fact the Sogdians has been engaged by Aegis, coming up in support of Erechtheis, and had broken, fleeing in all directions, largely to the south-west. Erechtheis had been almost destroyed by continual archery, however, and much to Athenian discomfiture, its remnants also made off. (At this point—the end of Period Fifteen—the situation is as shown in *Fig. 3.5*.)

Fighting had been prolonged and dusk was approaching—it had been arbitrarily decided that 20 periods should constitute a day, and the Persians had done much better than had been anticipated. However, at last, after no less than five periods of hand-to-hand combat, Parsa broke and fled, down to less than half its original strength, and the entire Persian right collapsed with it, unfavourable reaction tests taking unit after unit off in flight. The left, however, was very much in being, two more of the Athenian regiments having to fall back because of reactions and casualties. Thus when night fell only a few units of the Persian left were fightable, mainly the Mudrayan infantry, which had hardly been engaged, and the Lesbos hoplites, who had been moved at first to the right, then back towards the beach when the Persian right caved in.

The final situation is shown in *Fig. 3.6*—this being nightfall.

So, what can we make of the engagement? First, it was a tremendously exciting wargame, enjoyed by everyone, and the result a reasonable one, although by no means as clear cut a victory as the historical original. Indeed, out of strategical context it might be considered a drawn battle—actual figure losses being almost equal, 162 Persian figures and 168 Greek—but considered in relation to the campaign of which it was a part, it must be a Greek victory. The Persians had lost half their cavalry and altogether were in no condition to make any seaborne attempt against Athens. Indeed, their vaunted cavalry, although allowed to be a nuisance on the Greek right, which could have been avoided by an extension of that flank to the sea, did not affect Greek success on the centre and left. In the post-battle discussion it was argued that had the mounted troops been considered to have already embarked the Greek right would have advanced more or less as inexorably as the left. Maybe, but certainly, fighting the battle as we did, with virtually the entire Persian cavalry engaged, it seemed that even with their assistance, the Persians were no match for the Greeks—self-evident fact, in all conscience, I daresay—but it was pleasant to have our wargame rules confirming this.

IV THE IMMORTALS

THERE CAN BE little argument that, of the many celebrated corps to be found in the battle order of the armies of the Ancient World, the Immortals of the Great King of Kings of Persia merit a distinguished place, among such prestigious people as the Companions of Alexander the Great, the Theban Sacred Band and others of their ilk, and there can be few Achaemenid wargame armies not containing a proportion of these colourful fighting men. It is rather a pity, indeed, but nevertheless an historical fact that their recorded career was not a long one and that they had disappeared from the army of the Persian King when Alexander led his troops into Asia. A unit of Immortals is, in fact, to be found in the armies of the Sassanid kings many centuries later, playing a prominent role in the Battle of Daras (AD 530) against the famous Byzantine general Belisarius. The Sassanid Immortals were vastly different from their Achaemenid predecessors and in contrast to the latter were a cavalry unit probably made up of cataphracts or clibinarii, that is, armoured cavalrymen with horses wholly or partially armoured as well. Having thus briefly acknowledged the existence of these latterday mounted Immortals, we can dismiss them in favour of those with whom we are immediately concerned, the Household Troops of the King, the Great King, the Achaemenid!

Of these we read first in the pages of Herodotus, the ancient Greek historian, who gives us a vivid picture of the muster at Sardis in 480 BC of the army of Xerxes, King of Persia, this in preparation for his projected invasion of Greece to avenge the defeat suffered by his predecessor Darius at Marathon. It is no part of our intention to dwell on the enormous numbers of Xerxes' army or on some of the quite outlandish types contained therein, many from the most distant corners of the great Persian Empire, but we will note at the outset just what comprised the guard or household troops of the King. Herodotus describes the army's march and writes that the King was followed—among others—by 'a thousand spearmen . . . all

46

men of the best and noblest Persian blood . . . then a body of native (i.e. Persian) infantry ten thousand strong. Of these a thousand had golden pomegranates instead of spikes on the butt end of their spears, and marched in two sections, one ahead and one behind the other nine thousand, whose spears had silver pomegranates'. I think we should make a distinction, which will be referred to anon, between the first thousand and the ten thousand. It is the latter we hear of nest from Herodotus, who says that the 'Ten Thousand' were the first troops to cross one of the great bridges of boats which had been thrown across the Hellespont, being thus presumably the first Persian troops to set foot in Europe during this particular operation, and we read also that 'Xerxes himself with his spearmen'—these I suppose were the thousand—crossed during the second day. I feel it likely that the thousand were in fact the actual personal bodyguard of the King, the Ten Thousand forming a species of élite corps, a kind of *colonne infernale* of assault troops.

Herodotus provides further details, giving the names of the six generals who commanded 'all the infantry except the Ten Thousand', describing them as picked Persian troops commanded by one Hydarnes, and saying that the unit 'was known as The Immortals, because it was invariably kept up to strength; if a man was killed or fell ill, the vacancy he left was at once filled, so that the total strength of the corps was never less—and never more—than 10,000'. This seems to be in line with my suggestion above on the functions of the Immortals. Although, to judge from the descriptions of the historian, all the variegated national contingents in Xerxes' army were costumed most colourfully—or in some cases were notable for their lack of attire—it seems that the Immortals were clad in an especially brilliant manner. We can indeed get a splendid picture of them from the meticulous description of Herodotus. The Persians in general wore soft felt caps, a sleeved and embroidered tunic, armour resembling the 'scales of a fish', and trousers. They carried short spears, powerful bows and defended themselves with wicker shields. We are fortunate in having a splendid piece of monumental evidence to turn to for what we must assume to be the ceremonial garb of the Immortals, this being one of the glazed brick walls of the ancient palace of Susa. The élite Persian guardsman is shown in his characteristic dress— long, elaborately patterned and embroidered robe, a fillet of twisted cord round his head, 7-ft thrusting spear, the bow slung on the shoulder and a large quiver on his back. Hair and beard are elaborately curled.

It may be significant that neither side arms nor armour are shown, the lack of the latter being sometimes explained by the suggestion that the figures shown are on interior guard at the palace and would consequently not require armour. Another, and possibly a more valid idea is that the scale armour tunic referred to above was worn under the robe, this having some backing in that—if we may anticipate events—the Persian cavalry general, Masistius, when unhorsed at the Battle of Plataea a year later, could not be given his quietus at first because, as Herodotus writes, he was wearing 'a corslet of golden scales under his scarlet tunic'. It might well be that it was something of a fashion for generals and Household Troops to adopt this practice, for the Persians were always at pains to present a most dazzling appearance, Herodotus saying that every man glittered with the gold he carried about his person. It is certainly reasonable to assume that the rank-and-file Immortal preferred to wear his mail below his robe to allow the embroidery of the latter to be visible. This elaboration was, in passing, more than a little astonishing to the Greeks, who remarked upon it more than once, Plutarch (*Aristeides*) describing the 'embroidered clothing and good ornaments to cover soft bodies and faint hearts . . .' This last was, in the case of the Immortals, a totally unjustified sneer. It is doubtless true that a goodly proportion of the troops from subject races had to be urged into battle by their officers' whips, but the true Persian—and Mede as well—could fight with the greatest ferocity, as we shall see.

Anyway, proceeding with the story, we reach the point where the army of Xerxes, having marched through Thrace and north-eastern Greece, was brought to a halt by the Greeks—including the Spartans of Leonidas—at the Pass of Thermopylae. This, for the heavily armoured and splendidly equipped Greek hoplites, was a near perfect position, on one side steep cliffs and on the other the sea, the pass being about 30 yards wide only, or so it is said. The advantage of missile power was with the Persians—so many of them being bow armed—but in the circumstances this was not overwhelming by any means, a continuous line of Greek hoplites, with helmets, huge overlapping shields and greaves providing a virtually unbroken line of metal. After some delay Xerxes sent forward his excellent Mede and Cissian infantry, but their numerous and determined attacks throughout the day were beaten off, whereat their place was taken by Hydarnes and his Immortals. After hammering furiously at the Greeks they fell back, having had no more success than their predecessors.

If one reduces the situation to its simplest it seems that, basically, the principal reason for the Greek success was—as well as the armour —the comparative length of the spears used by the antagonists, the 9-ft spear of the Greeks being much superior in close action to the 6 or 7-ft one of the Persians. The latter suffered dreadfully, there can be no doubt, their superior numbers being of no avail in the confines

Above: Sherdan mercenary—possibly Pharaoh's guard.
Above right: A Nubian Auxiliary archer.
Right: An Egyptian heavy infantryman *(see Chapter I)*.

A Persian light cavalry-
man, armed with bow
and light spear or
javelin. The illustration
is derived from a figure
in the Iranian Publica-
tion, 'A Guide to the
2,500th Anniversary
Celebration Parade'.

A Persian bowman in action,
taken from the 'Guide to the
2,500th Anniversary
Celebration Parade'. The shooting position adopted is particularly interesting, and seems to
derive from a reference by
Xenophon ('Anabasis') when he
described archers encountered in
Kurdestan as using this
technique.

A Greek 'hoplite' or heavy infantryman, armoured, carrying long spear and short sword, and defending himself with the great, round shield, the 'hoplon'. Based upon Greek vase illustrations *(see Chapter III)*.

A Persian infantryman. Based upon a figure from a glazed brick wall at Susa, it is notable that the warrior, although armed with bow and spear, appears to have no armour, nor indeed a shield, although one possible explanation is that if engaged in palace guard duties, such articles were not carried *(see Chapter III)*.

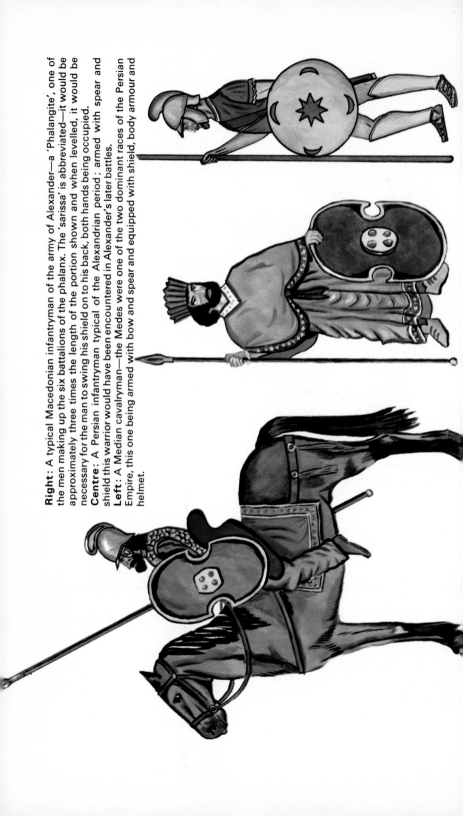

Right: A typical Macedonian infantryman of the army of Alexander—a 'Phalangite', one of the men making up the six battalions of the phalanx. The 'sarissa' is abbreviated—it would be approximately three times the length of the portion shown and when levelled, it would be necessary for the man to swing his shield on to his back, both hands being occupied.

Centre: A Persian infantryman typical of the Alexandrian period; armed with spear and shield this warrior would have been encountered in Alexander's later battles.

Left: A Median cavalryman—the Medes were one of the two dominant races of the Persian Empire, this one being armed with bow and spear and equipped with shield, body armour and helmet.

A Persian Immortal *(see Chapter IV)*.

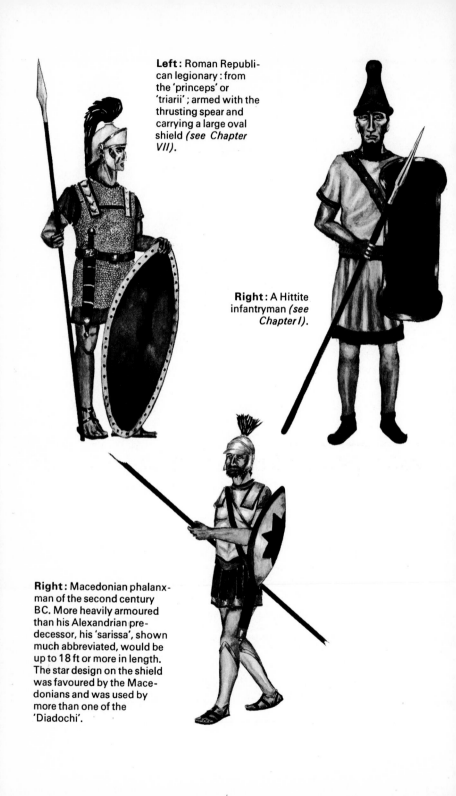

Left: Roman Republican legionary: from the 'princeps' or 'triarii'; armed with the thrusting spear and carrying a large oval shield *(see Chapter VII)*.

Right: A Hittite infantryman *(see Chapter I)*.

Right: Macedonian phalanxman of the second century BC. More heavily armoured than his Alexandrian predecessor, his 'sarissa', shown much abbreviated, would be up to 18 ft or more in length. The star design on the shield was favoured by the Macedonians and was used by more than one of the 'Diadochi'.

A later Thracian with helmet, 'thureos', greaves and 'machaira' —a reconstruction from literary and archeological sources *(see Chapter VI)*.

A fifth century Thracian with typical hat, cloak and a pair of spears (from a red-figured crater of the middle of the fifth century) *(see Chapter VI)*.

An early Thracian peltast with characteristic hat, crescent shaped shield, and high boots, about 550 BC *(see Chapter VI)*.

of the Pass. The 'Hot Gates' certainly earned their name on this first day of fighting, and even Herodotus has to admit, a trifle grudgingly, that 'the Spartans had their losses, too, but not many'. In any event, the Greek shield-wall held and the Persians drew off exhausted, as they were obliged to do again on the following day, after many hours of fighting.

Xerxes, as every schoolboy knows—or at least as he should—was extricated from his predicament by the traitor Ephialtes, who offered to guide the Persians through the mountains to outflank the Thermopylae position. Accordingly, led by him, Hydarnes and his Immortals marched all night through the mountains, brushed aside a flank guard of Phocians, and by first light had reached a position from which they could debouch upon the rear of Leonidas, again frontally attacked with the utmost fury by the Persians. All the allied Greeks had been sent off, leaving only the Spartans, and but few of them remained alive, holding a little hill to the rear of their original position, until they were overwhelmed by the Immortals from the rear and the main Persian army from the front. It had been a most desperate fight all round, and there seems no doubt that a very considerable 'intake' would have been required to sustain the Immortals at their traditional 10,000 strength.

We can pass very quickly over events immediately after Thermopylae—the occupation of Athens, the Battle of Salamis and so on. After this Greek naval victory, Xerxes thought it best to return to Asia, and shortly after the battle he left his general Mardonius in Thessaly with a powerful army, the aim being to complete the subjugation of Greece in the following year. Mardonius was allowed to pick his own troops for his army and not unnaturally he chose the Immortals, although their commander, Hydarnes, returned to Persia with his king. Thus it was that, in the army of Mardonius, the Immortals found themselves in the following year holding the left wing of the Persian army along the River Alopus, north of the town of Plataea, from whch the battle which followed took its name.

We need waste no time in detailing the preliminary events before the final battle, the manoeuvring, the cavalry attacks and so on, but rather come at once to the final morning when Mardonius' army advanced at the double in pursuit—as it thought—of the retreating Greeks, and the Immortals found themselves confronted by the Spartan contingent of the allied Greek host. As always, punctiliously religious, the Spartan king, Pausanias, awaited favourable sacrificial omens before attacking. During this delay, his men suffered very severe losses, the Persians having set up a barrier of shields from the shelter of which they poured a storm of arrows into the Greeks. Finally, the sacrificial omens were found to be propitious and the

Spartan phalanx charged grimly foward, first tearing down the shield wall then engaging the Persians hand to hand (it seems that dense masses of reserves crowding up close behind the front ranks of the Persians was a contributory factor towards reducing their fighting ability to say nothing of hampering manoeuvre).

The ensuing combat was one of extreme violence, of that there is no doubt, Herodotus describing it as bitter and protracted. It was far from being a walkover for what was certainly the finest heavy infantry Greece could produce, and even Herodotus is constrained to write 'Again and again the Persians would lay hold of the Spartan spears and break them; in courage and strength they were as good as their adversaries, but they were deficient in armour, untrained and greatly inferior in skill.' Plutarch (*Aristeides*) writes in like vein—'. . . the Persians fought bravely and skillfully before they fell. They seized the long spears of the Greeks with their bare hands, snapped many of them off, and then closed in to fierce hand-to-hand fighting, using their daggers and scimitars, tearing away their enemies' shields and grappling with them, and in this way, they held out for a long time.' However, at last the Persian general, Mardonius, riding a white horse, was struck down by a Spartan, at which the spirit seems to have gone out of the Persians, who fled to the shelter of a stockaded camp previously constructed in the rear. There they were able to hold out for a brief spell against the Spartans, who were without experience in assaulting field fortifications, but the Athenians and Tegeans were able to force an entry, after which it was simply complete slaughter. The bulk of the Persian army disintegrated and the Asiatic threat to Europe disappeared for a considerable time.

The mention of the shield barrier at Plataea raises some interesting points. The traditional Persian shield was apparently the 'gerrhon', such as can be seen carried by certain of the figures on a frieze in the Hall of the Hundred Columns at Persepolis. These figures are arranged vertically in several ranks, and only the first three of these guardsmen—for such they must be—carry the gerrhon plus a spear; the others carry only the spear and bow, the latter accompanied by different types of quivers, some the open ended type worn on the back as seen in the glazed brick walls at Susa. Some bowmen have, slung at the side, the large combination bowcase and quiver with cover.

All this seems to suggest that all the Immortals were not necessarily armed in the same fashion and it is a possibility that only the leading ranks—maybe three in number—were equipped with spear and shield for fending off attack, while succeeding ranks had the bow as their principal weapon and were employed largely in a missile role. With this in mind I may possibly be allowed to refer to *Fig. 4.1*, which shows one of the groups of Achaemenid infantry who marched past the Shah and many other notables during the military parade marking the 2,500 anniversary of the founding of the Persian Empire which was held in 1971. Now, and one presumes that this was based on historical research, we have the Immortals arrayed with the first three ranks armed with spear and shield—the former some 7-ft in length—and then two ranks of men armed solely with the bow, while these are followed by a number of ranks armed—if my eyes deceive me not—with bow and spear, but no shield.

This would seem to be a valid military formation, but one thing is something of a puzzle: I simply cannot see the shields as being suitable for setting up to provide anything like adequate protection for archers shooting from behind them. This leads me on to have a look at *Fig. 4.2*, also from the momentous parade to which reference has already been made. Here we have a front rank of infantrymen carrying a really large rectangular shield, about 5 ft × 3 ft in size and when carried as shown, affording a very considerable degree of protection, like the mediaeval pavise or the large Assyrian shield one sees in bas-reliefs. An illustration of an infantryman carrying such a shield is, incidentally, included in the illustrated guide to the 2,500th anniversary parade issued by the Iranian government.

Following the shield bearers in the photograph is a single rank of men without any shield but carrying a most interesting spear, much longer than the ordinary Persian variety and possibly about 9 ft overall, similar indeed to the long thrusting spear of the Greek hoplite. After this second rank come several ranks of more conventional infantrymen with the 7-ft spear and shield. The long spear is of particular interest, the suggestion being that its bearers took position behind the row of planted shields with their long spears projecting over the upper edge to dissuade any would-be attackers from coming too close. It would be probably difficult for more than one rank of spearmen to have their long spears projecting over such enormous shields, this, one supposes, being the reason for the single rank in the reconstruction of the Achaemenid infantry formation by the Iranian authorities. Should all this be the case it is easy to see how these large shields—of wicker, if we follow Herodotus—could be ripped away by the spears of the Spartan hoplites and would indeed, *in extremis*, be more of a hindrance in this context than a help.

I have in fact seen the suggestion that the ordinary gerrhon was hung on a pole—or a spear—stuck in the ground in front of the infantry, but in my view this is less likely for how does one manage to thrust the butt end of a spear, with a round ball thereon, into the ground?

Whatever might be the truth lying behind these fascinating speculations, as far as I know no more is heard of the Immortals after

Plataea, the Ten Thousand having no place in the Achaemenid forces
encountered by Alexander in his invasion of Persia. This was, of
of course, a century and a half later, but it seems likely that the losses
incurred by the unit in 479 BC where of such magnitude as to prohibit
the continuation of the Immortal 'replacement policy'. There was
probably not enough infantrymen of the quality required, this being
a problem which affected the Achaemenid dynasty throughout its
history.

Fig. 4.1

Fig. 4.2

Achaemenid infantry—*see text pages 50–51.*

V THE BATTLE OF THE GRANICUS 334 BC

Macedonians v Persians

ALL OF US have our favourite periods of history but certainly the Macedonian domination of the ancient world has many claims to be the most fascinating, partly by reason of the very colourful personalities involved, partly because of the number of striking developments in the art of war, these being largely due to Alexander the Great himself. Much of course was in his favour. He inherited from his father, King Philip of Macedon, a superbly trained and organised army and he succeeded to the throne when the power of Greece proper—Macedonia was deemed to be somewhat 'barbarous' and not part of that country—was declining, divided and enfeebled as it was through long internecine wars. At the same time his great enemy, the Persian Empire, was itself the victim of an inept system of government and had an army of the most vicious composition, relatively an agglomeration of the most diverse types of soldier possible. This is not to say, I hasten to point out, that the victories of the Macedonian army were easily gained, but of that more anon. It is undeniable, of course, that Alexander was immensely talented politically and militarily, and that he was sufficiently outstanding to come within the genius category is distinctly arguable. Wargamers certainly owe him a vote of thanks for the provision of several battles which can be examined and refought with the greatest of profit.

It can be said that, as well as his army, Alexander had inherited from his father an enemy, Persia, and it was to the destruction of this power that Alexander devoted his short life. His campaign was indeed a species of crusade against a nation which had previously invaded and devastated Greece and which held many of its people in thrall in what is now Asia Minor or Turkey. In actual fact Macedonia was not, as has already been noted, truly Greek but it comprised an aristocracy and a sturdy peasantry which provided, predictably, a first rate cavalry and a staunch and tough infantry. The Greeks, it is true to say, did not look upon the Macedonians with unequivocal friendship, and to ensure their cooperation or at least their neutrality, King Philip had already led an army against them, winning a decisive victory at Chaeronea in 338 BC. In the battle Alexander commanded the Macedonian left wing and contributed considerably to the victory. So, when Philip was assassinated in 336 BC the new king was already familiar with the problems of command as well as having a first rate and experienced army at his disposal.

It was far from being so with his Persian enemy, for the sprawling empire now ruled by Darius had declined in every possible way since the armies of Xerxes had hammered at the 'Hot Gates' of Thermopylae and, in spite of having fought fiercely, had suffered the defeats of Salamis and Plataea. A corrupt and venal political system, and the division of the country into 'satrapies', all combined to reduce the effectiveness of the Persian army, in which the emphasis was rather more on numbers than on quality.

To proceed, then. News of his father's death had been received with elation in Greece, but Alexander, by a whirlwind invasion stamped the seal of his authority upon that country, and he was in fact appointed Captain General for the war with Persia which was now an established aim. After subduing some recalcitrant peoples to the north of Macedonia and destroying the Thebans, who had again rebelled against him, he returned to Macedonia and in 334 BC crossed the Hellespont—the modern Dardanelles—with his main army, an advanced party having already established itself in enemy territory.

The Macedonian army was—like the Persian, and this is really the only point of resemblance—a pretty conglomerate sort of thing, less than half being true Macedonians. Many of the troops originated in Greece, as well as the Balkans, but it was a far more homogeneous entity than the Persians, and apart from its high fighting qualities its command structure was splendidly organised. In actual troop types it comprised everything from the heavy to the light, and its order of battle was based on two important elements, the phalanx and the heavy cavalry, the latter including the 'Companions'. Serving with them were the equally potent horsemen of Thessaly.

The phalanx consisted of 6 battalions of men armed with the 15-ft 'sarissa' or pike, and was organised generally in a 16-deep formation, enabling several rows of spear points to project beyond the front rank when in action. It required very smooth ground for successful operation—it was easy for such a huge and close-ordered mass to fall into disarray on rough or irregular terrain—and it was employed by Alexander to 'pin' the main body of an enemy while he manoeuvred his heavy cavalry to strike at a weak point. Alexander was usually to be found riding with the Companions who were well armoured, although they were at this stage without shields, and armed with a short stabbing spear. It should be pointed out that, at this period, stirrups did not exist and the spear was used in a free-hand stab, and not 'couched' beneath the arm in the manner of the latter-day knightly lance. Also forming an important part of the Alexandrian army were considerable numbers of Greek troops, largely hoplites, the traditional Greek heavy infantry, armed with a 9-ft spear, frequently armoured, and bearing, of course, the hoplon. These hoplites were either allied troops or mercenaries, and each type fought equally well. At the other end of the weight scale were numbers of light cavalry, Thracians, Paeonians and the like, including a regiment called the 'sarissophoroi'. These men were armed with what I think must have been a rather shorter version of the sarissa, not as long as the infantry weapon, but not the easiest thing to manage on horseback, as it required both hands to wield. In the ranks of the auxiliaries were archers from Crete—the legendary home of bowmen—and a unit of Agrianian light infantry, as expert with the javelin as were the Cretans with the bow. Finally we have certain troops whose weaponry and equipment still remain in some doubt, these in fact being the 'hypaspists'. Modern authorities fail to agree on just how they were armed or clothed, but what does seem certain is that they were of high morale, were able to take their place in the line of battle, but were nevertheless mobile enough to carry out rapid marches, cooperate with cavalry and so on. In all humility my own suggestion is that they were a lighter form of hoplite, armed with the 9-ft spear, but being without armour—or possibly having a leather or quilted linen cuirass—and carrying a lighter shield than the hoplon.

This, then, was the Macedonian array when, having advanced from the point where the crossing of the Hellespont had been made, it was confronted by the enemy on the opposite side of the River Granicus, which ran northwards into the Sea of Marmara.

This Persian host differed somewhat from those later encountered by the Macedonians, its composition varying in that it was a very hastily assembled force, apparently consisting entirely of Greek mercenary infantry and Persian cavalry. It was, in fact, the troops of the surrounding satrapies of Phrygia, Hellespontine-Phrygia and Lydia. One ancient writer, Arrian, states that it numbered 20,000 Greek mercenary infantry and a like number of Persian cavalry, but makes no mention of Persian foot, who presumably could not be mustered in time or whose numbers, if they were indeed present, were negligible. Major-General J. F. C. Fuller (*Generalship of Alexander the Great*) considers these figures to be grossly exaggerated and certainly Arrian did inflate Persian numbers in later Alexandrian battles, but the critic's suggestion that they should be cut to 5,000

Greeks and 10,000 Persians seems to be a trifle severe. Other writers have suggested some 15,000 for each, and this is the figure I have used —apart from anything else it does make the contending sides not too disproportionate, as we shall see.

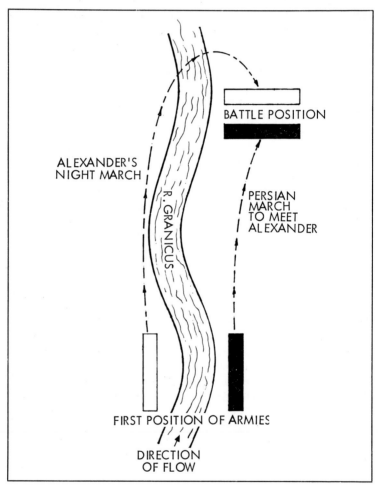

BATTLE POSITION

ALEXANDER'S
NIGHT MARCH

R. GRANICUS

PERSIAN
MARCH
TO MEET
ALEXANDER

FIRST POSITION OF ARMIES

DIRECTION
OF FLOW

Fig. 5.1

What is certain, however, is that on the further side of the River Granicus was deployed the Persian army, and it is at this point we are faced with one of the fascinating problems with which the researcher/wargamer has to deal. It must be said, initially, that there are two main sources for the wars of Alexander, these being the one already spoken of, to wit Arrian (Flavius Arrianus, who wrote in the second century AD), and Diodorus Siculus (whose account dates from the first century BC). Arrian, who is exactly followed by Major-General Fuller (his book was published in 1958), describes Alexander, delivering a direct frontal attack across the river immediately on his arrival, while Peter Green (*Alexander the Great*, 1970), argues that such a tactician as Alexander would never have carried out such an operation, and accepts the version of Diodorus, who says that the Macedonians halted, pitched camp and, during the night, marched downstream with great secrecy until a ford was found. Alexander crossed at this point, despite the efforts of a small force of Persian cavalry to hold him up, and formed up on the opposite bank, to await the enemy hurrying up to offer battle. In his much enlarged work published in paperback and entitled *Alexander of Macedon*, Professor Green devotes an entire chapter to the discussion, and after much fascinating speculation, comes to the conclusion that, after an abortive headlong attack, Alexander did, in fact, carry out the flank march and surprise river crossing postulated by Diodorus. (see *Fig. 5.1*). This version is, incidentally, accepted by another biographer, R. D. Milns (*Alexander the Great*, 1968). After considering the pros and cons of the two accounts, it was decided that our wargame should be based on Diodorus and that the battle should be fought on a more or less flat terrain—it was actually a gently rolling countryside, it appears—with the River Granicus on the right of the Macedonians and low hills to their left. The decision was something of a relief to the local 'Alexander' who viewed with some trepidation the prospect of getting his army across the river to meet—in an extremely disorganised condition—the immediate attack of the powerful Persian cavalry.

Let us again at this point refer to our modern authorities for the respective orders of battle for the opposing sides. These are almost identical and *Fig. 5.2* gives the actual dispositions of the two armies as they faced each other.

The Macedonians formed up in what was more or less their traditional deployment—Companion cavalry on the right, linked by the hypaspists to the six battalions of the Phalanx, and with Thessalian and Greek cavalry on the left wing. The front line of the Persian army consisted entirely of cavalry, with the infantry—the Greek mercenaries—forming a reserve to the rear. Little or no details of the composition of the Persian cavalry are available, other than their names—Median, Bactrian, Hyrkanian and so on—but it can be assumed that they varied from heavy to light and were mainly armed with javelins, although it seems pretty certain there would be some horse archers in their ranks. The best would be almost as good as the

Companions, but few would be of this high quality, although none lacked anything in the matter of courage. Their command structure is a trifle vague, and although Memnon of Rhodes, commanding the Greek mercenaries, was probably the most experienced soldier, there is little possibility that the Persian generals listened to his advice, let alone followed his instructions. In point of fact, in the order of battle Memnon appears to have had local command over the Greek cavalry—there could not have been many of them—on the extreme left of the Persian front rank, while the satraps and others, including Rheomithres, Arsites, Spithradates and Arsamenes, commanded large sections of the army.

The suggestion has already been made that we should take the Persian army to number approximately 15,000 infantry and the same number of cavalry, and it would seem appropriate at this stage to work out the strength of the Macedonians. It must at once be pointed out that a considerable part of Alexander's army which crossed the Hellespont a short time before does not figure in the battle plans illustrated by Fuller and Green. The missing men are the allied Greek, Balkan and mercenary infantry, amounting to no less than 19,000 men (cf. E. W. Marsden's *The Campaign of Gaugamela*, Liverpool Univ. Press, 1964, p. 38) and the total infantry present at the battle must therefore be only the phalanx battalions, the hypaspists and light infantry, amounting to little more than 13,000 men. The intriguing question is—where were the others? Possibly they were still moving up, or were in reserve some distance to the rear, or in fact it may be that they had been left behind at the first point of confrontation of the two armies across the Granicus and had not arrived in time for the opening of the action. All Alexander's cavalry were present, just over 5,000 men.

Before battle was joined, the Persians observed that Alexander—a most distinctive figure in elaborate armour and helmet apparently—had stationed himself on his right wing with the Companions, and deeming, quite correctly, that should a quietus be administered to the Macedonian, it would bring the battle to a premature end, they brought a considerable force of cavalry from their centre to their left. Upon which Alexander led forward his own cavalry at a great pace, first directly towards the Persian left—in a species of feint—but then swinging towards his own left, he crashed into the somewhat attenuated Persian centre. A most terrific hand to hand combat ensued round the person of Alexander himself, who was the obvious target of any number of glory-seeking Persians. First he was engaged by one Mithradates, a son-in-law of King Darius himself, but, striking him down, the Macedonian King was immediately assailed by a second Persian high-ranking officer, Rhoesaces, who succeeded

in landing a glancing, but severe blow on Alexander's head, fortunately encased in a metal helmet. Despite being considerably shaken
by this rude assault, Alexander struck down his attacker yet again,
but from his rear came a new onslaught, this time from the satrap
Spithradates. The latter was just about to strike with his axe when his
arm was smitten off by a tremendous slash from a lifelong friend of
the King, Cleitus the Black, doubtless saving his life. While all this
was going on the Persian right wing cavalry was being held in play by
Parmenio, commanding the Macedonian left and at the same time
the centre was steadily advancing. In face of this great pressure,
doubtless coupled with the serious effect on Persian morale of the
demise of so many senior commanders, both wings broke and fled
to be rapidly followed by the centre, and the entire front line was in
flight. There was left only the Greek mercenary infantry in the second
line. Attacked on all sides, they lost heavily and finally their remnants surrendered.

Alexander had won his first battle in Asia.

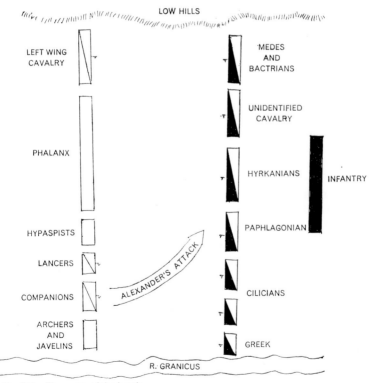

Fig. 5.2   Elements of the battle

The wargame refight

Two factors become immediately apparent during the process of re-creating an historical battle as a wargame. These affect the actual mechanics of the game and are first: the amount of room available for the encounter—this generally being a tabletop—and second: the number of troops, that is to say the actual wargame figures, which may be utilised. They have to be used with some care in conjunction with each other to produce as realistic and accurate a simulation as possible. Table size is obviously very relevant and the bigger available —within the limits of convenience in respect of the ability of players to manipulate figures in the centre—the better will be the game. Certainly if the battle is one wherein much manoeuvre is to be expected a large table area is essential and similarly, when a large scale engagement is being fought, one will have a much livelier 'feel' of having participated in a large battle if the number of figures deployed is such as to justify this. Against this, however, may be set the principle that if the battle is one where wide-ranging movement—say some sort of flank manoeuvre—is to be expected, then it is unwise to cram too many figures into the available space so that such operations become difficult if not impossible. For instance, in a battle such as Chancellorsville (1863) one would have to limit pretty strictly the space occupied by troops in the centre of the battle area to allow for the possible repetition of the wide flank march carried out by Stonewall Jackson, whereas at Malplaquet (1709) where the flanks were limited by woods, then the entire width of the playing area could be taken up by the lines of battle. The latter conditions, in a way happily, will operate for us at the Granicus, where the opposing armies faced each other across a fairly flat stretch of ground, bounded by the river on one side and a line of low hills on the other. So, as happened in a fair number of ancient battles, it was a headlong clash between the two armies, with little or no attempt at manoeuvre, on a wide scale at any rate.

Thus we were able to employ the full width of the table—9 ft, the other dimension being 7 ft—and the minimum of terrain features was necessary, involving only the river along one side and a suggestion of hills on the other. There was thus no clutter to impede movement, the battlefield consisting, as we have already pointed out, of gently rolling ground, ideal for cavalry and presupposing, in wargame terms, a perfectly flat playing surface. So, with the battlefield, such as it was, determined, it was time to consider the number of troops we were going to deploy, and to relate their numbers to those of the actual armies in the most appropriate possible way. Again, as we have noted, these were: Persian horse and foot 15,000 each, and Macedonians—13,000 infantry and 5,000 cavalry. In round figures then the ratio of Persian to Macedonian was about 15:10. In working out the number of figures to be used there is, incidentally, a further important consideration, this being the simple question just how many can be produced by the participants. This was easy for the Macedonians, but for the Persian cavalry host certain units bearing scant resemblance to the horsemen of Darius had to be pressed into service from the shelves, some being definitely akin to such distant peoples as Gauls and Germans. Not to worry, though, they sank their old national identities in the new and performed right nobly as well! In the actual determination of numbers I do not subscribe to the system which lays it down that one figure represents 100 actual men or whatever—it does not work all that well and my own preference is for organising the appropriate units, apportioning to each its correct space on the table, then filling up each unit's space with figures. This of course is far from being a rigid system, some juggling having inevitably to be done, but it gives a much more accurate representation of the manner in which different sections of the battle line functioned. In the present case, for instance, the 6 battalions of the Macedonian phalanx were reduced to 4, and more or less by the same token, for ease and flexibility of handling, the Companion cavalry was organised in two regiments. This, though, is really only a reflection of the truth as any large unit must consist of smaller sub-divisions, and certainly it was thus for the Companions. A very large body of troops on the wargame table while certainly providing an effective striking force becomes difficult to manoeuvre simply by reason of its size, particularly if it is cavalry. Sub-units are always useful to protect flanks and support the main body of the unit.

Anyhow, to proceed with the actual numerical details, the numbers in the two wargame armies were as follows:

The Macedonians

Infantry

1	Cretan archers		20
2	Agrianian javelinmen		20
6	Hypaspists		30
7	Phalanx: Battalion One		40
8	,,	Two	40
9	,,	Three	40
10	,,	Four	40
			230

Cavalry

3	First Companions	20
4	Second Companions	15
5	Sarissophoroi	20
11	Thessalians	20
12	Thracians	20

95

Total   325

The Persians

Infantry

L	Boeotian hoplites	60
M	Corinthian hoplites	60
K	Messenian javelinmen	35
N	Prodromoi javelinmen	35

90

Cavalry

A	Greek Light Cavalry One	25
B	Greek Light Cavalry Two	25
D	'Horsetail' cavalry	25
E	'Crescent' cavalry	25
F	Hyrkanians	35
G	Paphlagonian One	30
H	Paphlagonian Two	30
I	Median cavalry	30
J	Bactrians	25
C	Cilician cavalry	30

280

Total   470

(The figures and letters on the left will identify the units on the accompanying maps).

As to the weaponry, the phalanx, a 'medium' one with no body armour for the phalangites, was armed with the sarissa, of course, and it was decided that the hypaspists should be light/medium infantry equipped with the long, 9-ft thrusting spear plus a shield. With the exception of the Sarissophoroi and the Thracians the Macedonian cavalry were 'heavies' and it should be noted that the Companions, though armoured, had no shields.

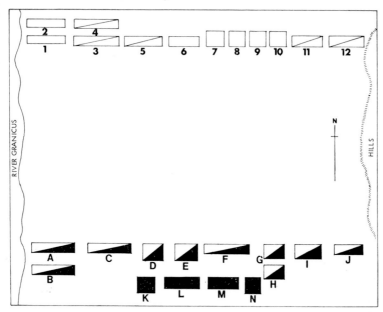

Fig. 5.3   Initial wargame situation

On the Persian side, the Greek mercenary infantry were standard medium hoplites, or light/medium javelinmen, while the Greek cavalry component was 'light'. Of the actual Persians, the Bactrians were 'extra-heavy'—i.e. armoured men on frontally armoured horses —while the Cilicians and Medians were 'heavy'. The remainder of the regiments were 'light'. In contrast to the Macedonians, all of whom had 'regular' status, many of the Persian mounted troops were 'irregular', an important factor when morale tests had to be taken.

Numbers, as we have already noted, were 325 and 470 for the Macedonians and Persians respectively, this giving a ratio of something like 10:14, a little lower than the actual one of 10:15.

So, the armies were drawn up as shown in *Fig. 5.3*, and a most splendid spectacle they provided, not least the Persians, some 280 cavalry figures in line being extremely impressive—it seemed occasionally to the participants that the table groaned under this ponderous mass of troops. On the subject of the participants, indeed, it might well be appropriate at this point to designate them and specify the roles they played in the drama about to be unfolded. Needless to say, they were all staunch members of the Society of Ancients and it is only fitting that the first to be considered is 'Alexander' himself, in the person of Ian Osborn (well entitled to this dis-

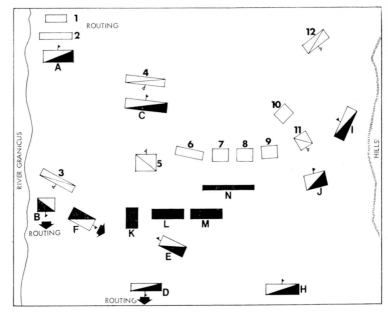

Fig. 5.4 Situation in Period Seven

tinction for at least the reason that he provided almost the entire Macedonian army). As was to be expected his miniature simulacrum took up its position with the Companions, commanding them and the right wing Macedonian light infantry. Next in line came Nigel Andrews, 'Nicanor', controlling the Hypaspists; Stanley Medrow as 'Perdiccas' had overall charge of the phalanx battalions, and Andrew Green, as 'Parmenio', handled the Thracian and Thessalian cavalry on the left wing.

While 'Alexander' appeared to have general command over his whole army, no single member of the Persian hierarchy had similar authority, and councils of war were frequent, but from the Persian left Barry Martin, as 'Memnon' looked after the interests of the Greek mercenary cavalry, 'Arsites'—Derek Casey, in fact—took over the left centre, David Matthews—'Spithradates'—had the right wing, and Peter Sheppard as 'Rheomithres' commanded the Greek infantry standing in reserve. Some liberties had, in fact, been taken with the Persian command, but let us proceed.

One important point to be noted before the dogs of war are unleashed is that the Ancient Rules of the War Games Research Group, Third Edition as amended, were used throughout, scoring was done by non-participants (thus absolving the players from this tedious chore) and there was an umpire to determine any situations where some sort of ambiguity had to be resolved. Finally, a word as to the orders as written by the commanders prior to the first movement. Those of 'Alexander' were a repetition of his illustrious prototype's, i.e. a diagonal attack across the table against the Persian centre by the Companions, with the Sarissophoroi supporting them, while the left wing held the powerful Persian right and the Phalanx and the Hypaspits advanced in the centre. The Persian scheme was that both wings should advance, destroy the enemy flanking troops and then close from both sides on the Macedonian centre. We shall see what transpired.

No time was wasted by either army, cavalry moving forward on all sides, and none more quickly than the Greek light cavalry of 'Memnon', which headed directly and swiftly (16-in. moves) for the Cretan archers. There is not much doubt that 'Alexander' was rather taken by surprise by this all-out thrust and the Cretans, without even having time to deploy, were charged by the leading cavalry regiment. They attempted to evade, but were caught in the rear and horribly cut up (this being Period Three.) They were driven through the Agrianians, who were able to hold out, the pursuing Greek cavalry being disorganised (on meeting fresh troops). Following up, the second Greek regiment was charged by the First Companions and speedily routed, in their flight nearly running into the Hyrkanians, coming up on their right rear. Had they done so, it would have been something of a disaster as the Hyrkanians, being 'barbarian' or irregular, would have been swept away in the rout. 'Arsites', however, had the foresight to move them out of harm's way just in time. Nor was this all the action in this quarter, the 'Horsetail' cavalry being engaged by the Sarissophoroi and also routed.

Meantime, Alexander's centre was advancing steadily and fierce fighting was taking place on the Macedonian left where 'Parmenio's' Thessalian and Thracian horse had to meet the onslaught of a mass of extra-heavy, heavy and light cavalry. They stood up to the ordeal very well, though, and one of the attacking Paphlagonian regiments was broken, although the heavy metal of the Medes was too much for the Thracians, who also fled. They were able to rally within a couple of periods, however, just before they went off the table, which would have put an end to them for good, of course. On the other side, the fleeing Paphlagonians were not so fortunate, disappearing into the blue, not to return. In pursuing them, however, the Thessalians became badly split and one isolated section was surrounded and destroyed by the Bactrian cataphracts.

While all this was going on, with the flanks of both armies hotly engaged, 'Perdiccas' was bringing his battalions steadily forward and on the Persian side 'Rheomithres' was moving his Greek infantry diagonally foward to his left to fill the gap left by the 'Horsetails', and also, it may be said, to avoid if possible a direct frontal attack by the very formidable Phalanx. We have at this point reached Period Seven, and *Fig. 5.4* shows the situation at this stage.

In spite of their success in routing two of the Persian left wing regiments, the Macedonians were feeling the strain and this was increased when the Second Companions were charged by the strong Cilician cavalry. This was the beginning of an epic struggle in which the Companions suffered no fewer than four consecutive 'push backs' (that is, they lost more casualties than their opponents in each of four successive periods but these were not sufficiently high to cause them to break). Thus, the next period was crucial—'A' Class troops, which the Companions were of course, break and run if pushed back five times in succession—and tension among the 'generals' was very high. However, a third force was about to throw its weight into the balance. This in fact was the Sarissophoroi, rallying back after their rout of the 'Horsetails', and they came pelting into the backs of the Cilicians. The ensuing encounter was a truly hectic one as, immediately prior to being smitten by the lancers, the Cilicians broke the Companions—having obtained the necessary fifth 'push back'—but, defenceless against this rude assault from the rear, the Cilicians were also broken, and they and the Companions went whirling off the table in one confused mass, leaving the Sarissophoroi to rally and cast about for a new opponent.

The tempo of events in the centre was also accelerating. The Prodromoi javelins were unwise enough to close with the 2nd battalion of the Phalanx and were driven off and, following up, the battalion collided with the Corinthian mercenaries and routed them (despite this the latter rallied a couple of periods later). However, 'Rheomithres' now brought up his Boeotians to cover the Corinthians and with them charged the advancing Hypaspists and, simultaneously putting in a skilful attack on them with the Messenian javelins, first pushed them back, then routed them.

It was now Period Twelve and the excitement was very high— perspiration flying about in every direction. (It was a very warm evening and ties, jackets and sweaters were successively discarded as the wargame temperature rose.)

The climax, however, was close at hand. On the Persian right, after a brief lull with units rallying and regrouping, the heavy cavalry went into action, the Medes and Bactrians of 'Arsites' charging 'Parmenio's' Thessalians, now supported by the newly rallied Thracians. The fighting was prolonged, first one side then the other being pushed back or held, the Persians cleverly avoiding the 4th Phalanx battalion which had been swung in their direction. The remaining Paphlagonian regiment also added its weight to the fray and the pressure finally proved too much for the Macedonians, who broke and fled.

The end was at hand. The Second Companions had been attacked by the combined Hyrkanian and Crescent cavalry and overwhelmed by weight of numbers. In the centre the Hypaspists had already been broken by the Boeotians and at nightfall (this being in fact the end of Period Fourteen the situation is as shown in *Fig. 5.5*, with the battalions of the Phalanx clubbed together and beginning to retreat towards the Granicus fords, covered by the Sarissophoroi. It seemed that, as far as we were concerned, Alexander's invasion of Persia had for the time being been brought to a halt. The wargame was over.

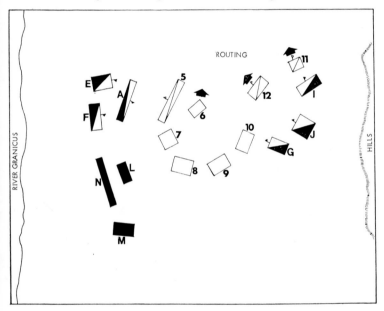

Fig. 5.5   Situation at the end of Period Fourteen

In retrospect it was enormously exciting and satisfying, despite its reversal of history and, of course, as on all such occasions the post-mortem was long and searching, being productive of far-ranging discussion. One point requiring immediate comment is that we allowed 14 periods for the day's fighting, rather longer than the ordinary quota but anything less seemed to be cutting the time rather unnaturally.

Space precludes anything more than the briefest mention of the principal points of this post-battle discussion, which mainly concerned the rules. The first related to the failure of the élite Companion cavalry, particularly in respect to their being pushed back five successive times until they broke before the Cilicians whom they had fought on nearly equal numerical terms. Now, the Ancient Rules of the Wargames Research Group, used for this reconstitution, although giving higher morale and greater powers of resistance (i.e. the four 'push backs' already mentioned) to élite troops, do not credit them with greater fighting potential or superior weapon handling capability. It is perfectly clear that, throughout history, there have been units—like Alexander's Companions—which, because of their greater combat effectiveness, experience, training and so on, were superior in overall fighting ability to the bulk of their contemporaries. We have the principle exemplified in the later horse and musket period when selected men were grouped together into grenadier units to be employed as assault troops, as was done by Napoleon and many other leaders. Accepting this proposition, then, it must be that in the ancient period there were similar bodies, of which the Companions are a prime example. What then has evolved in the group undertaking the battles of which I write is a system whereby any unit which has demonstrably greater fighting ability than its fellows, will be given a higher fighting factor than normal in a battle simulation in which it takes part. A similar procedure is followed in ordinary two-player competition games, wherein a submission may be made that a certain unit has historically greater combat effectiveness than is normal. This has to be proved to the satisfaction of the umpire or controller, and in the case of such approval the said body of troops receives a higher weapon factor. Another example of this would be Spartan hoplites in a Greek army and of course there are others. Incidentally, in the instance under discussion, the disparity was exaggerated by the Persian cavalry's having shields while the Companions were without. There is indeed a valid question in fact as to whether the Persian horsemen were really so equipped and here there is no little doubt. In truth, we have no monumental evidence to support their having shields, although, referring to a considerably earlier period, the historian Herodotus implies that Persian cavalry did have them and even earlier, there is a design on a bronze shield of the sixth or seventh century BC (found in Crete) which depicts mounted archers shooting at lions, the riders being shown carrying shields. This is reproduced in *Persia*, by R. Girshman (Thames and Hudson, 1964) with the comment that 'the motif could only have come from Iran', i.e. Persia. Again, Quintus Curtius (*History of Alexander*, IV, viii) writes of Darius' preparations for the Battle of

Gaugamela—three years after the Granicus fight—saying of the Persian cavalry that 'to those to whom he had before given nothing but javelins, shields and swords were added'. This, of course, is a trifle ambiguous, but it could be taken to imply that, earlier in the campaign and possibly at the time of the Granicus, *some* had shields and those who did not, received them in time for Gaugamela. Indeed, *The Illustrated Guide to the 2,500th Anniversary Parade*—the Parade was held in Iran in 1971—does in fact show Persian heavy cavalry with shields, so presumably the Iranian scholars responsible did feel that Achaemenid cavalry were so armed.

Another question arose as to the weapons used. The Alexandrian cavalry weapon was the 'Xyston', a stout thrusting spear some 6-ft in length, while that of the Persians was the shorter javelin (later re-placed by the spear) so that historically the Companions had a weapon superiority, but alas the Research Group Rules equate the spear with the javelin, giving the same weapon factor for each. Now it had been accepted before the game that the Research Group Rules as written should prevail as they do in Society championship games we play, but to obtain an accurate simulation of the action certain adaptations should have been made. However, quite inexplicably this was overlooked, this being something of a surprise as, in the past, special circumstances have merited extra value being given to certain troops. For example, when the Battle of Minden was refought some time ago, extra fighting qualities were given to the six British and three Hanoverian battalions which advanced against and de-feated the French cavalry. This is a logical proceeding for certainly, on that day at least, the troops involved had definitely very special qualities. Well, it is easy to be wise after the event, but had the Com-panions been given the attributes compatible with their historical prowess—and also to a slightly lesser extent the Hypaspists as well—the result might have been different, but who knows? The Persian effort must not be disparaged, and most certainly the whirlwind attack by the Greek light cavalry on Alexander's right really threw the Macedonians off balance, a circumstance from which they might not have recovered, irrespective of any other conditions. Still, all this considered, it was a splendid wargame, and it is no secret that 'Alex-ander' afterwards spent some licking his wounds and eagerly re-cruiting for his next trial of strength with the hosts of Persia.

VI THE THRACIANS

ONE OF THE earliest writers to mention Thrace and its inhabitants was the redoubtable Herodotus (*Book V*) and the picture he paints of them and of the practices to which they were prone is vivid in the extreme. Among some quite fascinating customs was their very vocal mourning on the occasion of the birth of a baby, the theory being that they were regretting the evils and mishaps to which the growing child would be subject in a far from perfect world. Certain of the Thracians also practised polygamy, and when a man died his wives strove to be acknowledged as his favourite, the successful claimant being forthwith slain and buried by her husband's side—it took all sorts, even in the ancient world! Apart from these interesting facts, what is of far more consequence is Herodotus' statement that 'if the Thracians could be united under a single ruler . . . they would be the most powerful nation on earth'. This is particularly interesting in that Herodotus wrote about the middle of the fifth century BC, when classic Greece and her component city-states were approaching their peak, although, despite defeats at Marathon, Salamis and Plataea, Persia must still be counted as the most powerful nation of the time. Alas for the Thracians, though, despite coming to notice with great frequency during the ensuing couple of centuries, this overall unification failed to materialise. The only occasion, in fact, when this seemed a possibility was in 429 BC, when one Sitalces was able to make himself overlord of most of Thrace and it seemed that, with the assistance of a variety of allied tribes, he might have created something of an empire, but he died without having been able to consolidate his domain. It might be as well to indicate at this stage that ancient Thrace was very extensive, covering present-day Bulgaria and European Turkey. With so much of this country of a hilly or mountainous nature, it is not really very surprising that—as with many peoples inhabiting such territory—the Thracians were of a warlike character, and it is with their exploits as such in the period from the fifth to the second century BC that we are at present concerned.

Of course, some hundreds of years earlier the Thracians had made known their name in a martial context, and Homer mentions their fighting as allies of the Trojans, although he fails to provide any details as to their weapons or tactics, a reference to chariots being simply in keeping with the 'heroic' warfare characteristic of those far-off days. It was not long, however, before the Thracians—the infantry, that is, with whom we shall deal at the moment—became firmly identified with a rather special type of fighting man, a kind of skirmisher who would ordinarily fight at long range with missiles, but who could, if occasion demanded, come together with his fellows into close order, and form part of a line of battle. Now, the usual missile man, slinger or archer, was not expected to do this, in the case of the latter and possibly also of the former, the absence of a shield for defensive purposes prohibiting such action. So there emerged the 'peltast', so called from the shield he carried—the 'pelta'—much lighter and smaller than the great 3-ft diameter one of the heavily armed infantryman, and it is almost entirely as peltasts that we know the Thracians, who, as the centuries passed, became pretty well exclusively identified as such.

We have to begin our narrative in 512 BC, when Darius the Great of Persia, seeking to create a foothold in Europe, crossed the Hellespont and invaded Thrace, thereafter marching north to the Danube, encountering on his way a number of the independent tribes making up the Thracian nation. Some of these surrendered forthwith to the King of Kings, but the principal tribe, the Getae, put up a stout resistance. They were, however, defeated and forced to join forces with Darius. After his campaign, the latter returned to Persia, but the adherence of at least a portion of the Thracians to his cause might have created something of a precedent for his successor Xerxes on his invasion of Greece in 480 BC.

It is to Herodotus, in fact, that we owe our first description of the Thracians embodied in Xerxes' army in that year. They wore, he writes, '. . . fox skins as head-dress, tunics with the "zeira" or long cloak, in this case brightly coloured, thrown over them, and high, fawnskin boots; their weapons were the javelin, light shield and small dagger'. These Thracians had, one supposes, submitted to Xerxes, as had several of the northern Greek states, the process being known as 'medising' (Greeks thought of the Asiatics as 'Medes' rather than as Persians). Apart from Herodotus' notes, however, we are fortunate in having a considerable body of evidence for the appearance of the Thracians from Greek vase paintings still preserved. They seem to have been a fairly popular subject for the vase artist, and almost half a century before the original incursion of Darius we have a cup (dated *c.* 550 BC) which bears upon it the

Fig. 6.1   Thrace and the neighbouring countries

likenesses of two fighting peltasts. This gives important details on their equipment, which consists of javelin and crescent shaped shield. Oddly enough, this shape has been frequently identified with the pelta, but this word, it should be pointed out, includes circular shields as well. One of these two figures on the vase is bareheaded, the other wears a tall, pointed hat, with pendant ear and neck pieces —not unlike the Scythian type. Not a few problems are posed by this hat—the 'alopekis'—for it in no way matches up to the 'fox skin' description of Herodotus, although it is possible that the fur is very close, or indeed the skin may have been tanned and was, in fact, a very soft species of leather. Another feature is that the alopekis is sometimes shown—details are never very clear—as having the ear-pieces tied up round the head, leaving only a long and broad flap covering the back of the head and neck. This is the manner of its wearing as depicted on another vessel, an amphora dated about 540 BC. This would be about the time when Thracian mercenaries were said to have been brought to Athens—reputed origin of the amphora —by the tyrant Pisistratos.

By the beginning of the fifth century BC, then, it seems that Thracians were becoming fairly well known in Greece, where naturally they were looked upon as barbarian, although there is nothing implicit in the description that their behaviour was such in a modern sense but simply that they were not Greek-speaking. Nevertheless, events did show that they could behave in a fairly uninhibited not to say ferocious manner. Let us, however, return to the army of Xerxes, whose forces are put by Herodotus at quite impossible numbers. He gives the contingent of Macedonians and Thracians as a preposterous 300,000. A more likely, though still suspiciously high figure would be 30,000, but however many of Xerxes' men emanated from Thrace, they did not come to notice during the invasion of Greece nor are they spoken of at the Battle of Plataea in 479 BC Herodotus does tell a good story of Xerxes retreat to Asia. On his march of invasion he had apparently left *en route* his royal chariot, a magnificent equipage. This had fallen into Thracian hands in some unexplained manner. Its fate is unknown, but it is unlikely that the tribesmen disgorged it. Even though many had served with the Persians, their natural inclinations did not prevent their severely harassing the forces of Artabazus on their way back to Asia via Byzantium after the defeat at Plataea.

In 429 BC, Sitalces, whom we have already briefly mentioned, decided upon something like a career of conquest and called up to his standard as many of the Thracian tribes as he could, including the Getae and others bordering on the Danube, together with certain hill tribes known as the Dii. The historian Thucydides offers an interesting comparison by saying that some joined him as mercenaries and some as allies. There is no doubt that at this time there was a strong possibility of Thrace's achieving the status of a great power, but after overrunning a considerable part of Macedonia, Sitalces decided for a variety of reasons to cut short the campaign and so led his army home. This was the last real manifestation of any national effort on the part of the Thracians and henceforwards we come across them in the main in other armies, being found during the next couple of centuries serving in many parts of Greece and the adjoining countries, almost invariably as peltasts.

In point of fact the first general properly to appreciate the use of the peltast in the warfare of his time was Demosthenes who, in 426 BC, saw a force of hoplites wiped out by a band of Aetolian javelin-men, whose missiles brought down the heavily armed infantrymen who could not catch their nimble adversaries. The successful employment of peltasts—possibly Messenian rather than Thracian—on the Island of Sphacteria remained in Demosthenes' mind and prior to his expedition to Sicily in 413 BC he recruited a body of 1,300 mercenary peltasts from Thrace (from the Dii, according to Thucydides). Unfortunately however, before the Thracians were able to reach Athens, Demosthenes had been obliged to sail without them and he, of course, came to a sad end in the ensuing campaign. In later fighting the generals Kleon and Brasidas made considerable use of peltasts, the former employing them in a variety of roles, including the provision of a small force to infiltrate a besieged city and open the gates to the attackers. Kleon also recruited Thracians and in 422 BC he met Brasidas at the Battle of Amphipolis in eastern Macedonia, on this occasion his army being mainly hoplites, while Brasidas had numbers of peltasts and cavalry. Kleon's troops were outmanoeuvred and fled pursued by peltasts, Kleon himself being slain by a Thracian peltast from Myrkinos.

It is only fair to point out that, on more than one occasion, Thracians lived up to their reputation for violence, which doubtless justified their being regarded as one of 'the lesser breeds without the law'. One such instance concerns the 1,300 peltasts who had arrived too late to join the Syracusan expedition of Demosthenes. Short of cash to pay them, the Athenian authorities sent them back to their homeland by sea. Their progress was a hectic one, culminating in their storming the town of Mycalessus in Boeotia, sacking it with the utmost thoroughness and slaughtering every soul in the place, irrespective of age, sex or condition. Thucydides, with no little justification, describes them as 'the most blood-thirsty barbarians'. The sequel is more interesting from our point of view, for the Thebans caught up with the looters on their way from Mycalessus, and attacked them with cavalry. With commendable honesty, Thucydides, who was obviously not partial to the Thracians, admits that they performed very stoutly against the Theban horsemen, putting 'up a good defence by adopting the tactics of their country, that is to say, by charging out in detachments and then falling back again'. In fact, including enthusiastic plunderers who remained in Mycalessus and were 'mopped up', the Thracians lost only 250 out of their original strength of 1,300.

Thucydides' description of the Thracians' tactics is most interesting and coincides perfectly with what one would imagine from their character and their weaponry. Armed with spear and shield—the pelta, of course—they would, if they so desired, form up in a fairly solid body, but would be able to send parties darting out to engage the enemy—either at hand to hand if these were light troops or with javelins as missile weapons if hoplites were involved. A word as to armament might not be inappropriate at this stage, in fact. We have already spoken briefly of the pelta—the light shield—which seems generally to have been of wickerwork, covered with hide or even, in some cases, with bronze sheet. When exactly the Thracian peltast adopted the helmet which ultimately became known as the 'Thracian' is difficult to establish, but this had without doubt appeared by the middle of the fifth century BC. They probably came into general use rather later, and certainly the vase paintings generally show Thracians as wearing the alopekis in one of its varieties. Again, from vase evidence there are strong indications that for offensive purposes the Thracian was armed with a pair of spears about 7 ft in length and fairly slender, such as could be used either for throwing—as a javelin—or for close fighting. There is one important and certainly interesting variety in armament which is well worth mentioning. Describing the campaign of the Thracian King Sitalces of whom we have already spoken, the historian Thucydides writes of the tribes he called to his standard thus—'He also summoned a number of the independent hill tribes who are armed with swords. These are called Dii, and most of them live on Mount Rhodope'. This might seem to infer that this particular tribe was not armed with the spear and were known especially for their use of the sword. The actual word used by the historian is "machaira", which is generally identified as a heavy, slashing sword with the cutting edge on the inside of a long, slightly curved blade, rather like the *kukri* of the Gurkhas, and a most deadly implement in close combat. It was later used by some of the troops of Alexander the Great and was indeed recommended by Xenophon. We shall have cause to return to this weapon later.

It is obviously quite impossible in the present context to do more than briefly indicate the warlike activities of the Thracians over the years and we can only note that they were engaged in the continual warfare in the Chersonese and elsewhere. It seems, too, that Greek peltasts were also making an appearance, indicating that the services of the much sought-after Thracians were becoming a trifle expensive. In any event, a force of them fought under Alkibiades in 409 BC, a campaign which included the siege of Byzantium, and they also appeared in the army of Cyrus in his Persian expedition in 401 BC, 800 Thracian peltasts being part of the contingent of Klearchos. There were also numerous other peltasts who, from their origins, might well have been described as Thracian also. At the Battle of Cunaxa, where Cyrus was slain virtually in his moment of victory, the largely Thracian peltasts fought with some distinction, skillfully allowing charging Persian cavalry to pass through their opened ranks while showering them with javelins. The Persian commander, Tissaphernes, opted to ride on through Cyrus' army rather than again face the peltasts by a directly retrograde move (Xenophon's *Anabasis*). After the battle, with their general dead, the Thracians, in the fashion of mercenaries, considered the situation, deemed it hopeless and some 300 transferred their allegiance to the Persian King. The larger proportion remained with the Greeks in their famous march to the sea during the initial stages of which they attempted to cover the march of the hoplites, no easy thing, their javelins being greatly outranged by Persian arrows. When Xenophon's men were passing through the rugged mountains of Asia Minor the Thracians formed part of his advanced guard, taking part in skirmishes far too numerous to detail. Some actions were quite considerable, however, notably against the Colchians, and the Thracians were as responsible as any of the other troops of the expedition in ensuring that the sea was finally reached. Here the Greek army broke up, one large body of hoplites deciding to raid several large villages in Bithynia, a district formerly colonised by Thracians, the heavy infantry on this occasion being very roughly handled by the Bithynian Thracians, both peltasts and cavalry. This treatment was such that after suffering severe losses the Greeks had to ask for a truce. They were rescued by another body of hoplites under Xenophon, after which they were doubtless happy to return to their homeland.

We have numerous references to Thracians in our literary sources for the first couple of decades or so of the fourth century BC, but these become less frequent as the century progressed. There seems to be no special reason for this, for all sorts of fighting continued in Thrace itself, but possibly the increasing number of locally raised peltasts— in the Greek city-states, that is—meant that fewer of the original Thracians were recruited. There were, indeed, thinking generals who attempted, with the success of the Thracian peltast in mind, to create an even more effective variety of this valuable troop type. Best known was the famous Iphicrates, an Athenian, who was in command of a force of peltasts in Athenian service in 393 BC. These, although nothing is definitely known of their origin, were most probably Thracians, and with them Iphicrates won startling victories over hoplites. He was responsible for certain changes, converting hoplites to a species of peltast by lightening the armour, lengthening the spear and replacing the heavy hoplite shield with a smaller one. Iphicrates and his troops are, alas, outside our immediate terms of reference, but the existence of his own special peltasts may be one other contributory reason for the lack of mention of Thracians in the latter half of the fourth century BC.

All this time, of course, we have been considering the Thracians as infantry, and certainly the majority of our sources considers them as dismounted troops. Several vase paintings, however, show them as being mounted, and frequently, following Greek artistic convention, the mythical Amazons are depicted in the Thracian garb of 'zeira', alopekis and high boots. Without doubt they fought as cavalry in their native country, but did not venture much beyond Thrace other than as the peltasts we have been considering—or not for some time, at least. The quality of horses available to them might have had some bearing on this, the point being considered by Professor J. K. Anderson (*Ancient Greek Horsemanship*, Univ. of California Press) who deems the Thracian horse to have been a compound of 'all possible faults'. He puts this down to the fact that Thrace had been scourged by numerous wars and invasions, whatever was good, in the equine sense, being stolen or destroyed. However, as the fourth century progresses we do find Thracian cavalry coming to notice.

This came about under the increasing Macedonian domination of the middle fourth century BC, King Philip of Macedon overcame the Thracians in several battles, with such results that he considered it impossible for them to be a nuisance in the future (Diodorus Siculus, *Book XVI*, 70). As we know, Philip was assassinated and it was his son, Alexander the Great, who led his armies into Asia, this being after he had subdued Greece and frightened the Thracian tribes into submission. We find, indeed, a contingent of Thracians included in his army. The only source to give a breakdown of Alexander's army —Diodorus (*XVII*, 70)—is a trifle vague, but 900 Thracian and Paeonian scouts (prodromoi) are specified, and also mentioned are 7,000 Odrysians, Triballians and Illyrians, the first two being Thracian tribes. Before the Battle of Issus (333 BC) we find Alexander using the 'light armed Thracians' to reconnoitre the mountainous surroundings of the Cilician Gates (Quintus Curtius, *IV*, 11) and at the subsequent battle the same source gives the Thracians in the van of the army, and 'these too were in light armour'. Necessity for speeding on with the narrative precludes more than the statement that during Alexander's advance through Persia the Thracians played an active part in the fighting and indeed were reinforced from their homeland on two occasions, 500 cavalry joining at Memphis in Egypt and a further 600 cavalry and 3,500 infantry at Babylon. Both cavalry and infantry took part in the Battle of Gaugamela. Some Thracians left to guard the baggage were attacked by Persian cavalry which had exploited a gap in Alexander's line and they had to be rescued by infantry sent back for the purpose. Although I have found no reference to Thracians being present at the Hydaspes battle, it would be indeed surprising if they did not participate therein.

After Alexander's death there was an immense amount of fighting between the Successors in which I imagine the Thracians played their part but we must perforce make a very considerable leap forward in time to what might be described as the last fling of the house of the legendary Antigonos One-Eye, possibly the most colourful of the 'Diadochi', and we find that his descendant, Philip V of Macedon made much use of Thracians in his wars against the rising power of the Roman Republic.

It might be as well at this point to note that, since the earliest historical appearance of the Thracians, their equipment had altered not a little, and as the traditional armour of the Greek infantryman had become lighter, so had that of the peltast become heavier. The

Thracian helmet would be more generally worn, and by this time the pelta had been replaced by the 'thureos'. The spear, however, was probably unchanged and it is recorded (Livy *XXXI*, 39) that, fighting under Philip against the Romans in some very wooded country both the men of the phalanx and 'The Thracians, too, were impeded by their spears, which were likewise of great length, among the branches which protruded in every direction'. This is interesting, for Livy who, like many Roman writers, is rather loose in his use of military terms, employs two different words for the spears of the phalanx and of the Thracians, 'hasta' for the first and 'rumpia' for the second. Whatever the rumpia was, it differed from the hasta (sarissa?) of the phalanx, but more of this anon. Anyhow, freed from the drain of the Punic Wars, Rome was enabled to concentrate against Philip, and the Macedonian suffered a complete defeat at Cynocephalae in 197 BC, when he had 2,000 Thracians in his army.

This defeat, however, did not prevent his maintaining an active opposition to Rome and it was believed that, with his connivance, a force of Romans under one Manlius was ambushed by some 10,000 Thracians from the Astii, Caeni, Maduateni and Coreli tribes, who plundered the Roman baggage train, retiring from the field with their loot when night came. This took place in 188 BC. Philip's successor, Perseus, also employed numerous Thracians, both horse and foot, in his army, 3,000 of them, for example, being part of the Macedonian army mustered at Citium in 171 BC, and Livy (*XLII*, 68) gives a lively description of their engaging a Roman force in the same year. He writes that the 'Thracians, like beasts of prey long held behind bars, charged so vigorously with a great shout upon the Roman right wing, the Italian cavalry, that these people, courageous by nature and through experience in war, were thrown into confusion . . .' Subsequently, after a successful fight, although Perseus failed to press home an advantage 'The Thracians returned, bearing with songs the heads of their enemies impaled on spears.' They were certainly a vigorous, if not entirely a civilised people.

And so we come to the final point in our account of the military history of the Thracians and it is indeed an intriguing one. It concerns Perseus' defeat by the Romans—in fact the final extinction of the Macedonian kingdom—by Aemilius Paullus the Roman at Pydna in 168 BC. The Greek historian Plutarch devoted one of his *Lives* to the conqueror and in his account of the battle he writes: 'First the Thracians advanaed . . . men of lofty stature, clad in tunics which showed black beneath the white and gleaming armour of their shields and greaves, and tossing high on their right shoulders battleaxes with heavy iron heads'.

The Loeb Classics translation gives the word 'battle-axes' for the original Greek 'rhomphaia' and this presents something of a puzzle for this weapon simply cannot be identified with certainty. One suggestion (P. Barker, *Armies of the Macedonian and Punic Wars*) is that the rhomphaia is the same as the curious implement used by Dacian tribesmen and depicted on Trajan's Column and elsewhere. This is a species of curved blade on the end of a stout shaft, and it is referred to by H. Russell Robinson (*Armour of Imperial Rome*), describing it as a 'falx' or sickle, but he makes no suggestion that it was used by any peoples other than the Dacians (The Dacian Wars lasted from AD 101 to 106). There is in fact no archaeological evidence to identify this falx with Plutarch's rhomphaia, and in the many representations of Thracians on Greek vases I have seen no signs of Thracian armament other than the spears already mentioned. More than one modern historian has written, usually very tentatively, of the rhomphaia. Snodgrass (*Arms and Armour of the Greeks*) admits that identification is extremely difficult, but thinks that it must be a large cutting weapon, possibly a kind of halberd, as it hung from the right shoulder, giving Plutarch as his reason for the last idea. I venture to think this is incorrect—translated from the Greek the actual words are 'brandishing from their right shoulders straight rhomphaias of heavy iron'—there is nothing about 'hanging' in the text. Professor Snodgrass also refers to Livy's describing—spoken of above—the Thracian weapon as the 'rumpia', a sort of spear. In *Thracian Peltasts and their Influence on Greek Warfare*, J. G. P. Best feels there is a strong possibility that the rhomphaia might have been a long Celtic sword borrowed from the Celtic invaders of Thrace in the third century BC. This may be the answer, but again there is no literary or archaeological confirmation, although in truth it is an attractive theory. In passing, it should be noted that many centuries after our period Anna Comnena (*The Alexiad*) described certain Byzantine guardsmen as being armed with the rhomphaia but this seems to be a whimsical use of an archaic word to describe the Scandinavian axe.

To proceed with our theorising, it seems that there are two possibilities to be considered. First, we know that Livy described the Thracians as being armed with the rumpia, which was long enough to impede its users in wooded country, and which the Loeb translator called a 'spear', as opposed to the 'lance' of the phalanx. Now, Livy died in Rome in AD 17, having written his monumental history. Plutarch lived from AD 45 to 120, visited Rome and was given high rank by Trajan. He was not, as far as is known, a military man, but it seems inconceivable that he did not have access to Livy's work when he was in Rome. Could it not have been, then, when describing the Thracians at Pydna in 168 BC ('*Aemilius Paullus*') he followed the

Roman historian and used 'rhomphaia' when the latter employed
'rumpia'? I am no philologist but there seems to be a definite similarity between the two words. In fact the Latin dictionary has no
definition of 'rumpia', but refers to 'rhomphaea'—a long, missile
weapon of barbarous nations'. Again, lexicographers are not necessarily military men, but there it is.

On the other hand, we have already noted that, according to
Thucydides, certain Thracian tribes were specially characterised as
sword users, presumably as a primary weapon, and that this was the
machaira. Was the rhomphaia then a possible variant of this? The
machaira is certainly slightly curved but its axis overall is pretty
straight. At any rate, there is another piece of evidence to be considered. We know that Thracians emigrated to Asia Minor in the
second millennium BC and that the Phrygians are of Thracian origin,
as are the Bithynians, who are continually referred to in the ancient
sources as Thracian Bithynians. Well, on the famous 'Darius Vase'
(in Naples Museum) there appears, standing behind the Great King
of Persia, what may be described as a Phrygian or Bithynian guard.
He wears a species of derivative of the Thracian headgear, has the
typical spears in the left hand, and bears on his right shoulder a
longer and slightly straighter version of the machaira. The resemblance to the original Thracian is striking, but again, evidence must be
deemed to be inconclusive.

One further piece of evidence may be adduced to assist the student
in determining the armament of the Thracian. Chronologically, it is
the latest of our references and dates from the great days of Ancient
Rome and the gladiators of, say, the second century AD, one of
whom was the Thracian, who was armed, according to Professor
Michael Grant (*Gladiators*, Weidenfeld and Nicholson, 1967) with 'a
curved scimitar ('sica') and a small square or round shield', plus the
traditional huge helmet and two heavy greaves. These details apply
to a small figurine of a Thracian gladiator in the British Museum,
which shows all the characteristics already described, including the
round shield and the 'sica'. The shield, incidentally, is shown as being
strapped to the gladiator's left arm, but other monumental evidence
indicates that it was also used with a single-handed grip, and my own
view is that, having regard to its size, it was often held in this fashion.
This is only an opinion, of course, and is really irrelevant to the
discussion, which I hasten to continue by pointing out the sica—or
whatever—wielded by the Thracian. It would appear to have a
curved blade some 18 in. in length, but as the British Museum
figurine is very small—some 5 in. in height—it is not possible to be
precise as to its cutting edges, whether one or two. From the manner
in which the weapon is held I incline to the belief that it was double-

edged, with the concave one sharpened almost to the hilt, to allow for an upwards, 'ripping' stroke—nasty, but there it is. It is not, in fact, vastly different from the weapon carried by the Persian King's guard on the Darius Vase (the latter is straighter overall). If one is not derived from the other, both might have had a common ancestor in the Thracian machaira of which Thucydides writes. The same word, in fact, is used by Xenophon (*Anabasis VI*, 1) to describe the sword used by the Thracians in the 'Ten Thousand'.

These are the pros and cons of a truly fascinating problem, and while having my own opinion on the rhomphaia, I leave the decision to the reader, having learned not to be doctrinaire on matters of ancient weaponry.

VIII BATTLE OF PYDNA
168 BC

Romans v Macedonians

THE DEATH, in 323 BC, of Alexander the Great, who had no viable
heir, resulted in the almost immediate dissolution of his vast and
conglomerate empire. In point of fact, signs of its breaking up had
been apparent before his demise, for even his personal reputation
and the power he exercised could not keep entirely subservient the
various governors he had appointed to outlying provinces and, in the
manner of the satraps of the overthrown and discredited Darius, they
too hungered for absolute personal power. With Alexander dead, it
did not take long for his leading generals to arrogate to themselves
large portions of his territory and between these Diadochi—the
Successors—there raged almost continuous warfare for the next half
century or so. This is a fascinating, if complex, period of ancient
history, but it is with only one of the sovereign states which arose
from Alexander's domains that we are immediately concerned. It is,

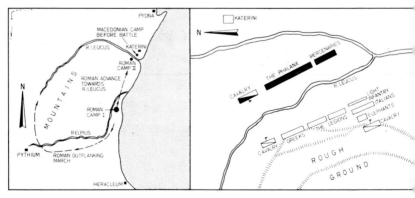

Fig. 7.1   Movements preceding the
battle

Fig. 7.2   Suggested orders of battle

in fact, that from which sprang the almost unparalleled example of human energy that Alexander had been and was, indeed, Macedonia. Initially seized by Antipater, he was succeeded by his son Cassander, whose reign was short and who died in 297 BC. His territory was immediately occupied by Demetrius Poliorcetes, son of Antigonos One-Eye—very much a larger than life character as were nearly all the Diadochi. Demetrius made a considerable name for himself as a general, but died in captivity after an unsuccessful campaign in Asia, leaving a son, Antigonos Gonatos—Knockknees—who, after sundry vicissitudes and not a little good luck, was able to establish himself as ruler of Macedonia. This was the beginning of a somewhat short-lived dynasty, which terminated with the battle which it is our purpose to discuss. Succeeding Macedonian monarchs had no easy time, however, being menaced by barbarians from the Balkans and having to engage in war with Sparta—now a mere shadow of its former self, it is true—and with various Greek confederacies.

It was with a rather more distant foe, however, that there was to be the terminal confrontation, for already in the third century BC the power of the Roman Republic was beginning to make its presence felt in the Eastern Mediterranean. Although at this time Roman efforts were more than sufficiently concerned with the wars with Carthage, it must have been evident that this rapidly rising power would ere long have much to do with every corner of the Middle Sea. Already victor in the first Punic War, which ended in 237 BC, Rome was probably the greatest power in Europe, although this fact was possibly not immediately apparent in the east. However, its potential authority was such that it was not long before its guidance and assistance were being sought by many a weaker nation. However, severe though the strain of the first Punic War had been the second proved markedly more desperate, and after the outbreak of fighting in 219 BC, Rome had many anxious days, most understandably with Hannibal's army scouring Italy and inflicting defeat after defeat upon the Roman army. The incredible resilience and powers of recovery of Rome came to the rescue and eventually Hannibal was obliged to quit Italy, being finally defeated at Zama in 202 BC. In a species of side shows to the Carthaginian operations Rome had already clashed with Macedon, some desultory and unimportant fighting taking place in Illyria. This so-called First Macedonian War ended in 205 BC, but seemingly unimportant though it might have been, Rome's memory was a long one, and a one-time enemy always remained an enemy, this despite Philip V's (grandson of Antigonos Gonatos) ostentatiously turning his attention eastwards and away from Rome. Alas for his hopes, Roman aspirations already included Egypt, in whose affairs Philip sought to meddle. Caught up in his machinations the Island of Rhodes appealed to Rome against the Macedonian predator. The Romans lost no time in declaring war against Philip and in 198 BC they had an army in the field in Greece, the local Achaean League (Greeks never seemed to have lost their innate animosity towards their northern neighbours) joining forces with the Romans. During the following year the Consul Flaminius, with his legionary army, met and defeated Philip's Macedonian one—based upon the traditional sarissa-armed phalanx—at Cynocephalae. The battlefield was rough and unsuitable for manoeuvring the unwieldy phalanx, the right and left wings of which became apparently separated. The left was charged and defeated by Flaminius in person, his legionaries being supported by elephants. At the same time an enterprising tribune made an attack against the rear of the other Macedonian wing and Philip, with the remnants of his army, fled the field with some urgency.

This was a very significant encounter and deserved to be better known than it is, as it was the first occasion when the powerful phalanx had been met and unequivocally overthrown by Roman arms. It could be said that the handwriting was on the wall for what remained of the Alexandrian military system.

Fortunately for the defeated Philip, however, Rome did not exploit the victory as it might have done, being still pretty well committed elsewhere. In the event, Philip still failed to appreciate the potential power Rome could exercise in the Eastern Mediterranean and he continued to interfere in the turgid politics of this unsettled area, despite his lengthy experience in such matters, completely misjudging Rome's intentions. He seems in fact to have come to the conclusion that the power of Rome was waning rather than waxing and with marked rashness, seized certain cities on the coast of Thrace. A Roman commission sent to enquire into this decided that Philip should withdraw his garrisons from these cities. He did so, apparently quite readily, but secretly began to prepare for war, even sending his younger son Demetrius to Rome as an ambassador, presumably to provide some sort of political smokescreen. On his return, his elder brother, Perseus, was so inflamed with jealousy at the younger man's apparently successful mission that he induced their father, Philip, to have Demetrius poisoned as a potential usurper. This he did, but the king's conscience was so affected by this dreadful action that he died two years later, in 179 BC, leaving his surviving son, Perseus, to be king of Macedon.

This individual was not by even the wildest stretch of the imagination cast in the heroic mould. As we have seen, he was both treacherous and inordinately jealous. Pusillinamity, vacillation and greed were also strong features of his character and, all in all, he had little of his ancestors, Antigonos One-Eye and Demetrius, in him and no one, indeed, contributed more to his final extinction than he did himself. Be that as it may, his devious plans concerning Greece and the neighbouring states disturbed the Roman senate, and after sundry comings and goings war broke out in 172 BC, preparations being immediately made by the Romans to carry the fighting into Greece. At this time, though, the situation was much to the advantage of Perseus; his treasury was full, he had great stores of food and a powerful army. It is just possible that a determined attack in the early stages of this Third Macedonian War might well have, for some years at least, held back the Roman tide, but he delayed and let the opportunity slip away. In addition, his behaviour towards possible friends was crass in the extreme. During the ensuing two or three years he alienated many powerful allies, whose armies plus his own would have outnumbered the Romans many times over. He made the most fulsome promises of the monies he would pay these people but his indubitable avarice at the last moment prevented his fulfilling these undertakings, and the potential adherents to his cause—including a Celtic chief who offered him 20,000 horse and foot—turned from him in disgust. However, even without such help, Perseus, more by good luck than sound tactics, and by the incompetence of the Roman generals sent against him was initially able to maintain his position and even to inflict the occasional defeat upon his enemy. Nemesis was fast approaching, however, this being in the person of Lucius Aemilius Paullus, who, as one of the Consuls for the year—it was 168 BC in present day reckoning—was entrusted with the command of the Roman expeditionary force. He was an experienced soldier, sixty years of age and of high reputation for forthrightness and honesty. His actions in Macedonia would serve to enhance his standing and right at the outset he lost no time in assembling his army and setting it in motion for the scene of operations.

His army, which it will be appropriate to consider at this stage had as a basis two legions. At this time their establishment was considerably less than the later imperial legion and the basic organisation was dissimilar. In battle, a legion was formed up in three lines, each consisting of ten maniples. Of these the first two lines had maniples of 120 men each, those of the third line being 60 each. The first line—the 'hastati'—presumably had pride of place and possibly incorporated the wealthier—and consequently the better armed—citizens, who paraded in mail tunics, wore helmet and greaves and bore large oval shields. The most important feature of the equipment of the 'hastatus' was the pair of 'pila' he carried, together with a short stabbing sword. As the legion came to close quarters they would be hurled at the enemy at very short range—possibly less than thirty paces. If the pilum did not hit and disable an enemy, it might well lodge in his shield, greatly inhibiting the actions of the carrier and allowing the hastatus to close up to him and get to work with his sword. Both second and third lines of the legion—the 'princeps' and 'triarii' respectively—were less well provided for and their main weapon, in lieu of the pilum, was a fairly long thrusting spear—7 ft or more in length. Their protection was usually in the form, not of the mail tunic, but of a square of bronze worn on the chest attached to a leathern shirt, although doubtless many would be clad identically to their comrades of the hastati.

In battle order the legion was drawn up with intervals between the maniples of each line, but with those of the princeps and triarii 'offset' so that each interval was covered by a maniple of the line behind. 'Chequerboard' is a fair sort of description of the system. In addition to these basic types, each line had a small attached force of light infantry—'velites'—together with a few hundred light cavalry. At full strength the establishment of the Republican legion was just short of 4,000 men, but as always this was doubtless much reduced in active service conditions. The Roman historian, Livy (one of our principal sources for the period) does say that each of the two legions allocated to Aemilius were reinforced to 5,200 infantry, but his account is generally vague and he gives no clue to the numbers actually mustered for the day of battle. In addition to the legions, however, 7,000 Latin allies—i.e. from other parts of Italy—plus 400 cavalry were enrolled and arrangements were made for 600 Gaulish cavalry to be sent from their homeland to the theatre of war. They are not mentioned later in any account, but there is no reason why they should not have arrived and taken part in the fighting. Light troops were always something of a problem for the Romans, but they did have a force of Cretan archers (they had been sent to Macedonia two years earlier—see Livy, *XLIII*) but it seems that at least as many of these sought-after missile men were serving with Perseus. As well as the Gaulish cavalry already spoken of many Numidian horsemen were serving with Aemilius, probably under the command of Misagenes, son of the famous Masinissa. From Numidia there came also a force of elephants, in number between 30 and 40 (two contingents of these animals had been sent, according to Livy). These troops then formed the Roman army, together with possibly six or seven thousand local allied Greeks, but numbers are generally hard to come by. Two vital leaves of the original Livy MSS are missing, these possibly containing battle 'states' and so on; Polybius would certainly have given the essential details but only fragments of his narrative exist for this period, and Plutarch (*Life of Aemilius*) writes a particularly vivid account, but he was not a military man, and his story, although splendid reading, lacks much of the data we would find useful. We have to make the suggestion, for reasons which will be plain in a moment, that the Romans mustered for the coming battle about 25,000 men, of whom about 3,000 were cavalry.

It was this array, then, which found itself on the banks of the River Elpeus in north-eastern Greece in the late spring of 168 BC. To the north was the camp of King Perseus of Macedon.

The King had a carefully chosen and well prepared position and an initial Roman suggestion that a frontal attack be made upon it was at once discounted by Aemilius. Its defences—garnished as they were with engines—were far too strong and the Roman general decided upon a policy of winkling the enemy out by a stratagem. To this end he sent a considerable force—over 8,000 infantry and some 200 cavalry—to Heraclium on the coast (see *Fig. 7.1* for the general situation) to suggest to the watching Macedonians that an amphibious operation against their rear was to be attempted. This decoy force, however, wheeled about and, marching at night, made all possible speed in the other direction with the object of outflanking the Macedonians not from the sea but from the land. The column marched as far as Pythium, then swung north-east to take the enemy in the rear. Alas, however, a Roman deserter—stated to be a Cretan—was able to make his way to the Macedonian camp and Perseus, apprised of the Roman plan, took action accordingly. Under one Milo, a strong force of some 12,000 men was sent to block the approach road and a lively encounter ensued, decidedly to the advantage of the Romans, with Milo and his men falling back in considerable disorder, which might in fact have been tantamount to a rout, towards the main Macedonian army. After this, with no little haste, Perseus moved his army northwards, taking up a position in the neighbourhood of Katerini, a village south of Pydna, on a fairly level plain which was ideally suited for the operations of the phalanx. Although two small rivers ran through this terrain, they were small and, at this time of the year, could have been only of small account.

Details of the strength and composition of the Macedonian army, as for the Roman, are sadly lacking, and it can only be pieced together and that withal in a pretty speculative fashion. Certainly, however, it was built around the massive sarissa-armed phalanx, this weapon by this time being rather longer than that used by Alexander's original phalanx, being as much as 18 ft in length and possibly even more. The sarissa, while obviously very effective and indeed well nigh irresistible when the phalanx with which it was armed was well closed up, was nevertheless unwieldy in the extreme and any sort of flank movement was virtually impossible, especially when the sarissas were lowered. It was very easy to make it fall into confusion and indeed Livy (*XLIV*) goes as far as to say that any sort of uproar coming from a flank was enough to disorganise it completely. Thus its flanks had to be very well protected, this being done by light or medium infantry. In addition to the hardcore of Macedonians, Thracians, Paeonians and others of their ilk made up the army, and a considerable part thereof, above all the troops furnished by the Greek allies, would be the traditional hoplite, by this time likely to be in leather armour instead of the old bronze. (The term 'hoplite' may be used very loosely here, but I employ it to describe Greek close order infantry armed with spear and shield. The monument to Aemilius Paullus shows a fallen Macedonian with a large, round 'dished' shield of, apparently, some 3-ft diameter, which seems to merit the term 'hoplon' and to justify the word 'hoplite' in this context.) On the contrary, the second century BC phalangite had metal armour, the weight of which doubtless added to his relative immobility. One section of the phalanx is referred to as the 'Bronze Shields'. And of course we have the Cretan archers serving with the Macedonians, and their cavalry was also fairly numerous, armoured and in contrast to the Companions of former years, armed with large round shields. Perseus had ample funds at his disposal and the mercenary section of his army must have been large—10,000 such troops had made up the greater part of Milo's force and the survivors of that expedition together with what remained with the main Macedonian army must certainly have amounted to a similar number. One extremely interesting unit is referred to by one ancient author— this being the anti-elephant corps, a most exotic body, it would appear, trained to attack elephants and having spikes on their helmets and projecting from their shields. The picture of these people rushing forwards and butting at the pachyderms' legs with their spiked helmets is a fascinating one, but, sad to say, they did not make their presence felt in the coming battle.

So, albeit unsatisfactorily for our purpose, we have to make some attempt to deduce the numbers of the combatants on each side and about the only statement we can make with any degree of confidence is that they must have been approximately—and that is the operative word—of equal strength, as there is no mention in contemporary accounts of any overlapping flanks. I suggest, therefore, that with the two legions amounting to about 8,000, plus an equal number of the Latin allies, together with possibly 5,000 light infantry, the Roman army amounted to about 21,000 infantry and some 3,000 cavalry, these being the Gauls and Numidians. I have to repeat that these figures are to a great extent conjectural and are of course liable to correction.

It is recounted (Livy *XLIV*, 11, 9–11) that during the previous year, Perseus had detached for certain operations 10,000 light infantry and 12,000 Macedonians, while he himself had troops remaining with him. Even if the last mentioned were merely body-guards this still makes a possible total of something like 25,000 men; so I tend to think of Perseus's army as being 12,000 Macedonians—I look upon these as forming the actual phalanx—plus 10,000 mercenaries and light troops and about 3,000 cavalry, although these would contain a fairly high proportion of heavy troops as opposed to the Roman light horsemen. Thus again we have about 25,000, roughly the same as what has been suggested for the army of Aemilius. And of course the latter had his elephants—thirty or forty as we have already estimated.

These then were the armies preparing themselves for battle in their respective camps on the plain we have spoken of, initially separated by the River Leucus. The Romans were some way up the slopes of Mount Olocrus, the low, rough hill to the immediate west of the river, while the Macedonians remained on the level ground, the essential condition for the operation of their phalanx. There was an eclipse of the moon on the night of June 21st/22nd 168 BC and we can therefore date the battle exactly—the following day in fact.

It seems that fighting began rather accidentally (at least according to Plutarch), a stray horse being pursued by a couple of Romans, who were set upon by Thracian pickets from the Macedonian army. More Romans came up to their comrades' assistance and ere long both armies were in battle array facing each other as fresh units issued forth from their camps, with the Romans deployed on the slopes of the hill where they were able to manoeuvre with some ease because of their open formations. As we have said the orders of battle must be pretty speculative, but J. F. C. Fuller (*Decisive Battles of the Western World*, Vol. 1) gives a very plausible layout of the positions of the armies. *Fig. 7.2* in general agrees with it but much must be of necessity inspired guesswork. It seems that the Macedonians advanced to attack very rapidly, much more so than the Romans anticipated, especially considering the normal deliberate progress of the phalanx, and it was the Roman right wing, behind which at some distance were the elephants and some reserve cavalry, which was first contacted. The Romans here fought stoutly—they seem to have been Pelignians under one Salonius—and drove back their opposite numbers until the clashed with fresh enemy, which may in fact have been the left hand elements of the advancing phalanx. This was too heavy metal for the Pelignians who were flung back in some disorder. They were pursued by Macedonian light infantry who in their own advance came up against the right-hand Roman legion—probably commanded by Aemilius in person—and their pursuit came to an abrupt halt. About this time the left-hand legion was ordered forward —under Lucius Albinus, an ex-consul—against the 'white shield' phalanx in the centre of the enemy line (Livy refers to two phalanxes —the White Shields and the Bronze Shields). By this time it seems that the battle, in the centre certainly, had been drifting westwards and even on to the lower slopes of Mount Olocrus, for the phalanx was showing imminent signs of fragmentation. Seeing this, Aemilius ordered his legionaries to exploit these rapidly appearing gaps, which they at once proceeded to do—once legion and phalanx became inter-mingled the long sarissa was useless when opposed by expertly wielded sword and shield. Simultaneously Aemilius brought up the elephants and the allied cavalry forming his follow-up force on his right and it seems that at this time the Macedonians began to break. The elephants were supported by Italian allied troops, presumably rallied after the initial reverse at the hands of the Macedonian left wing. However, notwithstanding the pressure, the Macedonians in the centre fought stoutly but the second legion's charge was too much and they fled, suffering horribly from the pursuing Romans. Perseus led the flight, accompanied by his own cavalry guard and the cavalry of the right wing. The slaughter was prodigious, thousands of Macedonians being slain; the King made good his escape, but was some time later taken prisoner on the island of Samothrace.

So ended this highly significant struggle between the military systems of Rome and Macedon and although the phalanx was employed some sixty years later by Mithridates of Pontus, as an effective fighting force it had had its day, and it was the legions of Rome which were to dominate the European military scene for many centuries to come.

The wargame refight

One fortunate feature about most ancient battles is that they were generally fought on reasonably level ground, with a conspicuous absence of villages or other built-up areas. Hence the wargame terrain for such an encounter quite often needs little more than a level table, and to this general rule Pydna is no exception in that all that was required to reconstitute the battlefield was the suggestion of foothills running up to Mount Olocrus (this being constructed of $\frac{1}{2}$-in. contour boards) and just a short stretch of the River Leucus—the latter more for appearance than strict necessity, as the troops were set out at the time when both armies were drawn up in order of battle, the Romans on the lower slopes of the hill and with the Macedonians having left their camp and crossed the river. Thus the battlefield is as shown on the various maps. In connection with this, however, one point should be made, this concerning the roughness of the slope. On this, it seemed, the Romans were perfectly at ease but which broke up, it appears, the tight 'dressing' of the Macedonian phalanx, to its ultimate discomfiture. The question was whether to make the hill of such a degree of irregularity that troops moving upon it would be 'disorganised' or whether it should be considered smooth. This was indeed something of a knotty problem as, should this be done, it would disorganise both Roman and Macedonian infantry (War Games Research Group Rules were again in use, by the way). As demonstrated by history, this would have been unreal, so, keeping fingers firmly crossed, we decided to leave the hill as a smooth slope and to hope that events would not show that some special rules should have been drawn up to differentiate between manoeuvre on the hill of legion and phalanx. Again, purely for the sake of aesthetics a few trees were placed upon the upper contour—the 'look of the thing' being considered to be not without importance. It was decided, too, that the 9-ft length of the table would provide a longer battle line, and consequently larger and more impressive armies, so this was done, allowing, when the armies were set out, more than 4 ft between the front lines (overall width being 7 ft).

Having set out the terrain, simple as it was, it was time to deploy the armies, beginning with the Roman, which was drawn up along the entire length of the lower contour. The difficulty of accurately establishing the numbers of the Roman army has already been pointed out, but in general terms it appeared that the two legions of Aemilius occupied about a third of the length of the entire line, at the centre, of course, and this seemed to be a convenient yardstick. One problem was at once apparent, this being how to represent the tripartite tactical organisation of the said legions—hastati, princeps and triarii. It was obviously essential to reproduce this system as far as possible in the wargame army and at the same time to avoid over-fragmentation of the legionary units, so, after some fairly prolonged discussion as to ways and means, it was decided, rather as a matter of expediency, to reduce the three line formation to two, the first being armed with the pilum, the second with the thrusting spear. The decision was influenced, not unduly but to a limited extent, by the number of wargame figures available, and with this in mind, it was determined to split the legionary component into three rather than the historic two. Thus we found ourselves with three legionary units, each comprising two bodies of 40 figures, that is, the hastati and a composite force made up of princeps and triaii. It might have been more realistic to have made the second (numerically) stronger than the first but this was not done, largely for the sake of maintaining some sort of homogeneity of the units. In this way the Roman centre consisted of what we called—and why not?—three legions, I, II and III, each subdivided, as for example 1A and 1B, these being hastati and princeps/triarii respectively and each legion totalling 80 figures.

As has been indicated, the Roman right wing was largely composed of Italian allies, probably of less value than the legions and there was also some cavalry. 2 units of infantry—Pelignians and Samnites—made up the former, together with Cretan archers and Numidian cavalry. The elephants posed something of a problem but the conclusion was finally arrived at that three would be an appropriate number to represent the historic 30 or 40. The left wing consisted of Greek infantry—hoplites—light medium javelins and Gallic cavalry, a small unit of 10 only, plus a few local archers. 'Aemilius Paullus' therefore deployed for action the following:

Romans

D	Legion IA	40
A	Legion IB	40
E	Legion IIA	40
B	Legion IIB	40
F	Legion IIIA	40
C	Legion IIIB	40
G	Pelignian infantry	20
H	Samnite infantry	20
J	Cretan archers	20
P	Numidian cavalry	10
K	Greek hoplites	40
M	Greek javelinmen	20
N	Greek archers	20
O	Gallic cavalry	10
R	Elephants	3

Total 403

That is to say, a total of 380 infantry, 20 cavalry, and 3 elephants (the letters in front of the units will serve to identify them on the maps). It will not escape notice that the native Romans are more numerous than they were in actual fact, but it might well have been that the Italian allies were armed as were the legions in which case the right hand legion—I—might be counted as 'allied' rather than as 'Roman', but the point did not seem to be important enough to warrant any alteration in nomenclature.

Set out to occupy a frontage roughly equal to the Roman, the Macedonian array was naturally built around the phalanx, this after some consideration being of five battalions, each of 30 phalangites, making up a total striking force of 150 men, narrower in frontage than the opposing legions but, in wargame terms fighting four ranks deep in contrast to the one and a half ranks of the legionaries, it was a formidable force, certainly to be reckoned with. Its left was immediately supported by a strong body of allied Greek hoplites, the Achaeans, 40 in number, as well as by Lycaean archers, together with light medium infantry—called, for want of a better term, the hypaspists. Thessalian and Thracian cavalry formed the extreme left, the latter being sarissophoroi (almost certainly incorrectly, but they did provide something different). On the right of the phalanx were two more units of Greek hoplites with archers in support, and two cavalry units took their place on the extreme right, plus some javelinmen. The Macedonian battle order was as follows:

Macedonians

1	Phalanx—Battalion I	30
2	Phalanx—Battalion II	30
3	Phalanx—Battalion III	30
4	Phalanx—Battalion IV	30
5	Phalanx—Battalion V	30
9	Achaean hoplites	40
10	Lycaean archers	20
11	Hypaspists	30
14	Thracian cavalry	10
15	Thessalian cavalry	10
16	Peloponnesian hoplites	40
6	Boeotian hoplites	40
7	Pella archers	20
12	Greek cavalry I	10
13	Greek cavalry II	10
8	Macedonian javelins	20

Again we have an army of 400 men (the number in front of a unit indicates its map position, of course). It will be noticed that the Macedonians have 40 cavalry to the Romans' 20. There is no real justification for this, other than the feeling that, with Macedon and north-eastern Greece to draw upon, it seemed not unlikely that the Macedonians would be able to deploy considerably more cavalry than the Romans, whose local allies would probably be infantry. That was the theory, at any rate.

So, with these two armies in mind, we observe how they were deployed for battle in *Fig. 7.3* the Romans on the lower slope of Mount

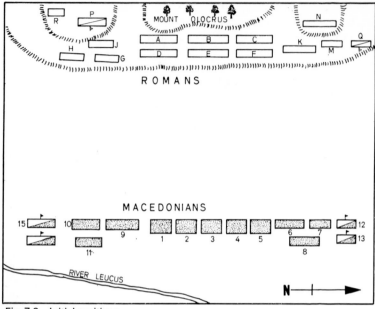

Fig. 7.3   Initial positions

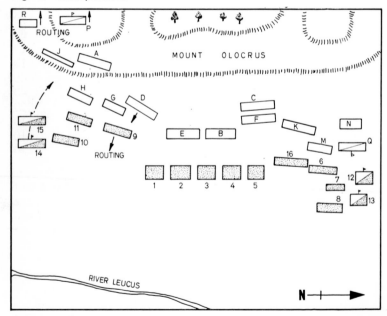

Fig. 7.4   Period Five

Olocrus and the Macedonians, having crossed the trickle—relatively speaking—of the river Leucus, but let us leave them contemplating each other for a moment while we briefly refer to the actual wargame command elements of the two sides. Following normal practice, players, as they arrived, drew a slip of paper with the name of the historical commander whose function they were to assume, and, for the Romans, the following took over the mantles of authority. 'Aemilius Paullus' was Derek Casey, and he took command of the three legions as well as having general authority over the entire army. Alan Stoneman—Scipio Nasica'—had the left wing, while Mark Smith became 'Fabius Maximus', having command of the Roman left as well as 'doubling' as 'elephant arch' of the three massive animals to his right rear. Similarly, the role of the egregious King Perseus fell to Andrew Green while David Matthews as 'Milo' held the Macedonian right and Peter Sheppard had the left as 'Evander' (this individual's name is not recorded in the battle accounts but is later referred to as a companion of the fleeing Perseus, so he was pressed into service, on the assumption that he must have had some place in the battle).

It can now be revealed that the battle plans adopted by the two sides were fairly similar. The centre of each army was to advance and engage the enemy, while each left wing was to drive ahead, the main objective of the Macedonians being to deal as quickly as possible with the Roman elephants—at long range if possible—while the Roman left was to prevent the Macedonian cavalry from threatening the flank of the phalanx. As we shall see these plans resulted in a sort of clockwise movement of the battle fronts.

So much for the preamble, battle may now commence.

At once both armies advanced, the Romans moving down to the level ground; at the same time the Macedonian left wing cavalry, with the Lycaean archers in close support, drove rapidly ahead and as early as the first period, arrows were flying to and fro, the bowmen of both sides taking casualties at extreme range. Seeing the approach of the strong left wing Macedonian horse, 'Fabius' brought forward his elephants in the hope that they would neutralise the approaching Thracians and Thessalians. Alas, as soon as they came within cavalry charge range and were thereby obliged to test their morale, the elephants promptly stampeded, i.e. went into the prescribed two periods of uncontrolled advance. Naturally enough, the Macedonian cavalry duly evaded and, when the elephants came to a halt, archery fire again imposed another reaction test upon them, this second one being a complete reversal of the first, the animals incontinently fleeing (one cannot do much with a three dice throw of 4!). Sadly enough for 'Fabius', resulting reaction to nearby units caused him yet another pang, his Numidian cavalry joining the elephants and making their way off the table by the shortest route! He was thus left with the Cretan archers and the two Italian allied units. All this rather anticipates events, though, but it serves to dispose of the Roman right for the time, although much was transpiring elsewhere. (It was on the fifth period, in fact, that the elephants were finally 'written off'.)

In actual fact fighting moved right along the line from the Roman right, both Roman javelin units being engaged by the hypaspists and the Achaean hoplites, the latter pushing back the Pelignians (history thus repeating itself). Legions and phalanx were steadily coming closer to each other, but so far in the northern sector there had been only exchanges of archery, both 'Nasica' and his Macedonian counterpart, 'Milo', manoeuvring with some caution. Event followed event on the Roman right, though, with the Thracian cavalry—the Sarissophoroi—backed by the Thessalians, attempting to move round the Roman flank. They were dissuaded, however, by the appearance of Legion IB, detached from the centre. It appeared now that the Roman commanders determined to deal with the Macedonian left before the legions were fully occupied by the phalanx, and to the aid of the Pelignians, now pushed back for the second time, there came Legion IA, which charged the Achaeans, first pushing them back, then breaking them completely (the legendary 'free hack' as the unfortunate Greeks turned to flee was colossal—256 casualties 13 figures, in fact; no doubt about the effectiveness of the pilum armed Roman legionary. *Fig. 7.4* shows us the position at this point—Period Five.

By this time, further to the north, the Peloponnesian hoplites had advanced some distance beyond the phalanx line, and were immediately charged by Legion IIIA, being forthwith broken. At the same time coming up on the left flank of the IIIA, Legion IIIB collided with the Boeotian hoplites, pushed them back, then broke them. So far, the Romans had had much the better of the fighting but there was a long way to go yet.

With the Numidian cavalry and the elephants now out of the way, part of the Macedonian left wing cavalry, the Thracians, swung inwards, in an attempt to charge the Cretan archers, but coming under heavy archery from their intended victims, had to have a morale reaction—it was a bad one—'halt two periods'. The Thessalians were more fortunate. Making a sweep behind the Samnites, now engaged with the hypaspists, they drove home against Legion IA, just engaged with the 1st battalion of the phalanx. This double impact was too much for the Romans, who broke and fled, pursued by the cavalry. Fortunately for the nearby Romans who might have

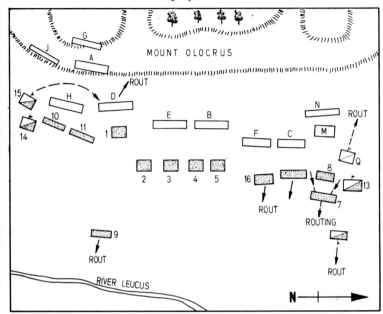

Fig. 7.5   Period Twelve

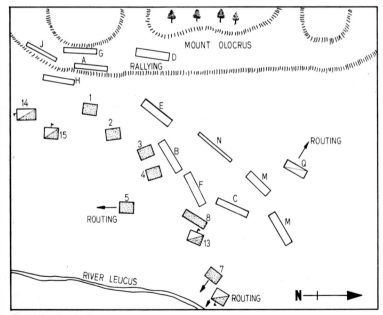

Fig. 7.6   The End

been affected by this catastrophe, their morale tests were all 'good'. Right at the other end of the line there had been some to-and-fro action. First, Greek light cavalry 1 had been broken by the Roman cavalry—the Gauls—but they, in turn, had been routed by the second Macedonian Greek cavalry regiment, but this, fearing to be outflanked by the Greek javelinmen (M) and suffering from fire from the Greek archers (N) did not follow up, rallying back for further action. The Macedonian javelinmen (7) were at this stage in two minds whether to go to the support of the phalanx or to move to the support of the right flank. 'Perseus' and 'Milo' were not altogether sure of the best course, as by this time (Period Eleven) the centres of both armies were close to collision. However, their minds were made up for them when 'Nasica' pushed his own javelins rapidly forward to engage their Greek opponents, bringing off a devastating charge, to break them and send them back in rout to the rear. Fortunately for the Macedonians, their cavalry regiment (13) had rallied, and thus threatened, the Roman unit fell back without making further advance. *Fig. 7.5* shows the situation at this point.)

By this time, however, the centre of the battlefield, which had been slow to show much activity, now began to spring to life. Possibly both 'Aemilius' and 'Perseus' had been a trifle apprehensive of the power of the phalanx and the legions respectively, but by now they were locked in combat. It was the 5th battalion of the phalanx which fiirst experienced the crunch. It was attacked in front by Legion IIB and in flank by IIIA, now rallied from its rout of the Peloponnesian hoplites. Not unexpectedly, this double assault was too much for the battalion and it forthwith broke, pursued by IIIA. Having partially connected with the 4th phalanx battalion in its charge, IIB remained to fight with this unit, while IIA took on both 2nd and 3rd battalions. The 1st battalion had moved forward into the position earlier occupied by the routed Legion IA and was as a result a trifle isolated. The routed legion had in fact rallied and was poised to renew the combat. Near at hand, the Sarissophoroi had suffered from Cretan archery and had had first a 'retreat three periods' reaction, this being immediately followed by another which was 'rout' and the lancers vanished from the table to the south. At the same time, their Thessalian comrades had also had a 'halt two periods' reaction. With the Cretan archers on the hill, together with the rallied Pelignians and Legion IB, together with the largely untouched Samnites, facing it, what remained of the Macedonian left was in a most doubtful position and appeared to await events in the centre. The other Macedonian wing was not in a materially better situation, the Pella archers falling back before the Roman javelinmen, while the Greek cavalry on the Macedonian side was under fire from the enemy archers. All hinged now on the phalanx as the day was rapidly coming to an end—Period Fifteen.

In fact, however, it was now obvious to the Macedonian command that the battle had been lost. The 4th phalanx battalion had been pushed back twice by Legion IIB, its flank was threatened, and an 'Army Reaction' was imminent. 'Perseus' now ordered a general retreat and conceded that the Roman had won the day. It was assumed that, as the stipulated 'day' had run its course, darkness would inhibit pursuit and the Romans would not follow up. *Fig. 7.6* shows the position at the end of the fighting.

This then was our reconstruction of Pydna and, apart from noting that all enjoyed it enormously, no special comments are really necessary. The Research Group Rules worked efficiently and dice throws for reactions and weapon factors generally showed nothing that was wildly improbable, apart from the unfortunate ones for the elephants. It would have been interesting had some provision been made for a right or left stampede as well as the forward or backward type, but that will have to be discussed later. There was one minor dissatisfaction concerning the employment of the Roman legionary formation. As the narrative shows, the hastati and princeps/triarii sections were used as separate units. This did not seem to be quite right and there was some discussion concerning how to devise a more accurate representation of the Republican three line formation. The problem was to avoid the use of a large number of really miniscule units which would have been the case had every single maniple been represented by only a few figures. Eventually, after much discussion, it was resolved that possibly the best simulation of the Republican legionary system was to retain the three line concept—hastati, princeps and triarii,—and to have the first two consist of three units, each of 8 figures, and to have two 6-figure units as the third. This retained the numerical ration of the three lines—2:2:1—and the total of figures in the legion—60—would not be unreasonably inflated and indeed adhered to the requirement of the Wargames Research Group Rules, which stipulate 60 figures (including subunits) as the maximum permissible.

APPENDIX 1–BIBLIOGRAPHY

IN COMPILING a bibliography for some work he has prepared, there is a tendency for an author to assemble under this heading a vast catalogue of books which may have some reference to his subject, and which, he may possibly hope, will impress the reader with his erudition. This is a trap into which I do not propose to fall. The volumes I list below will be the absolute minimum, and all of them I have found of use in my study of ancient warfare, whether they be the products of modern scholarship or the original sources. Of the two categories, the latter is of supreme importance. They are generally not difficult to obtain, many being in translation in the admirable Penguin Classics series, although a small word of caution may be in order, few if any of the translators being military men and their translations are couched in a very free style. If possible, therefore, it would be as well for the reader to use the Loeb Classics series for his study, with parallel Latin/English and Greek/English texts. These splendid little volumes, now, alas, costing more than double the price I paid for the first I acquired, are invaluable and include many most informative notes. First, then, I list the ancient sources which roughly cover the period appropriate to this volume (there are others somewhat more obscure but those mentioned will suffice for a beginning):

Arrian	Polybius	Dio Cassius
Diodorus Siculus	Quintus Curtius	Thucydides
Herodotus	Strabo	Xenophon
Plutarch	Livy	

Next follow the modern works containing the most relevant material for both historian and wargamer.

Adcock, F. E. *The Greek and Macedonian Art of War*, 1957
— *The Roman Art of War under the Republic*, 1940

Anderson, K. J. *Military Theory and Practice in the Age of Xenophon*, 1970
—*Ancient Greek Horsemanship*, 1961
Best, J. G. P. *Thracian Peltasts and their Influence on Greek Warfare*, 1969
Burn, A. R. *Persia and the Greeks*, 1962
The Cambridge *Ancient History*
Fuller, J. F. C. *The Generalship of Alexander the Great*, 1958
—*Decisive Battles of the Western World*, Vol. I, 1939
Green, P. *Alexander of Macedon*, 1974
—*The Year of Salamis*, 1970
Hignett, C. *Xerxes' Invasion of Greece*, 1963
Kontorlis, K. P. *The Battle of Marathon*, 1973
Marsden, E. W. *The Campaign of Gaugamela*, 1964
Milns, R. D. *Alexander the Great*, 1968
Russell, H. *Armour of Imperial Rome*
Rossi, L. *Trajan's Column and the Dacian Wars*, 1971
Tarn, W. W. *Alexander the Great*, 1948

To the above it is essential to add several of the Wargames Research Group's publication which provide, by way of text and line drawing, a vast amount of information on the armies which have been met in this book. Although originally designed as inexpensive handbooks, prices of the more recent have been possibly higher than the term 'handbook' might suggest, but they are nevertheless of the greatest possible value and can only be unreservedly recommended—as of course are the ancient period wargame rules produced by the same group. The relevant titles are:

Buttery, A. *Armies and Enemies of Ancient Egypt and Assyria*.
Nelson, R. *Armies of the Greek and Persian Wars*.
Barker, P. *Armies of the Macedonian and Punic Wars*.

The published literature on wargaming in the ancient period includes *War Games through the Ages 3000 BC to 1500 AD* by Don Featherstone (Stanley Paul, 1972), *Ancient Wargaming* by Phil Barker (Patrick Stephens, 1975) and I should be less than human to omit my own *The Ancient Wargame* (A. and C. Black, 1974).

Finally, there can be no doubt that it is essential for the 'ancient' wargamer to become a member of the Society of Ancients, which concerns itself with wargaming in the period and which produces a bi-monthly journal, *Slingshot*. The Subscription is £3·50 annually and the Treasurer, to whom monies should be sent, is Malcolm Woolgar, 44 Shaftesbury Avenue, Worthing, Sussex. I can only say that Society membership is a 'must' and is indeed well worth the subscription cost.

APPENDIX 2–WARGAME FIGURES

ONE UNEXPECTED feature—to me, certainly—of the flood of new wargame figures which has in the last few years inundated the market is the fact that many retailers have simply been unable to stock the quite staggering variety of every period being produced by an increasing number of manufacturers, several having decided that they have not the room to cope with anything like a representative cross-section of what is available. Consequently, the wargamer, unless very fortunate with his local model shop may find it quicker to make his purchases direct from the manufacturers, of whom the principal producers of ancient period miniatures can be found in the following list.

Warrior Miniatures, 23 Grove Road, Leighton Buzzard, Bedfordshire.
Lamming Miniatures, 45 Wenlock Street, Hull.
'Tradition', 5A and B, Shepherd Street, London, W.1.
Miniature Figurines, 28–32, Northam Road, Southampton.
Hinchliffe Models Ltd., Meltham, Huddersfield, Yorks.
Greenwood and Ball Ltd., 61 Westbury Street, Thornaby on Tees, Teesside.
Dixon's Miniatures, Ash Grove, 17 Royles Head Road, Longwood, Huddersfield, W. Yorks.

All the above bring out the 25 mm figure, which is the most popular size, but 15 mm types are also in use, produced by Miniature Figurines as above, as well as Peter Laing, Minden, Sutton St. Nicholas, Hereford.

GLOSSARY

Alopekis—tall pointed hat with pendant ear and neck pieces, worn by Thracians.

Antilabe—Continuous sequence of leather or rope loops round the inside of the 'hoplon'.

Boeotian Helmet—wide brimmed metal hat with brim bent down at sides originating in Boeotia.

Companions—Alexander the Great's household cavalry.

Composite corselet—leather tunic—sometimes covered with metal scales—laced in front, and with shoulder pieces.

Cuirass—armoured protection for the body, metal or leather with metal plates, generally consisting of back and breast protection.

Diadochi—The 'Successors'—Alexander the Great's generals who set up their own kingdoms after his death.

Epomides—Shoulder pieces fixed to the back of a composite cuirass, drawn over the shoulders and tied down to the chest in front.

Falx—A weapon used by the Dacians, a curved blade fixed to a stout, wooden handle.

Gerrhon—A Persian shield, probably of wicker work, some 3 ft long, oval shaped with semi-circular cut-outs on each side.

Hasta—The Roman Spear.

Hastati—the front line units of a Roman Republican Legion.

Hoplite—The classic Greek heavy infantryman, armed with spear and shield—the 'hoplon', hence the name.

Hoplon—Circular, 'dished' shield about 3 ft in diameter, carried by Greek infantrymen.

Hypaspist—Elite infantry of Alexander the Great; possibly a lighter version of the hoplite.

Khepesh—Sickle type of sword used by the ancient Egyptians.

Kopis—General term for the heavy sword used as a side arm by the Greek hoplite.

Machaira—Probably the heavy, curved slashing sword used initially by the Thracians and later by the Greeks.

Maniple—subdivision of the Roman Republican legion, either 120 or 60 men.

Na'arun—Believed to be 'recruits', a unit playing a part in the Battle of Kadesh.

Peltast—An infantryman, more lightly armed and equipped than the hoplite in classical Greece, frequently a Thracian.

Pila—Very heavy javelins with long iron heads sunk deeply into stout wooden shafts. Carried by the hastati.

Pilos—originally the felt lining of a Spartan or Athenian helmet, later made of bronze and used as a helmet *per se*.

Polis—the city state of ancient Greece.

Porpax—Straight bar with armband, across inside of the hoplon.

Princeps—Second line units of a Republican Roman legion.

Pteruges—Species of kilt made of pendant strips of stout leather.

Rhomphaia—A weapon—very difficult to identify—but associated with the Thracians.

Rumpia—Latin word probably identical with the Romphaia.

Sarissa—A pike, varying between 15 and 21 ft in length.

Sarissophoroi—Light cavalry in the army of Alexander the Great believed to have been armed with a shorter version of the sarissa.

Satrapies—political divisions of Persia in the fifth and fourth centuries BC.

Sica—a dagger or short sword.

Spolas—Tunic of leather or canvas worn by various troops of ancient Greece.

Successors—see Diadochi.

Ten Thousand—see list

Thureos—A substantial oval shield with a central boss and sometimes a rib along its length.

Triarii—Third line of a Roman Republican Legion.

Velites—Light infantry attached to a Roman Republican Legion.

Zeira—A long and heavy cloak, (usually brightly patterned worn by Thracians.

INDEX

INDEX

NOTES